CDMA Systems Capacity Engineering

For a listing of recent titles in the *Artech House Mobile Communications Series*,
turn to the back of this book.

CDMA Systems Capacity Engineering

Kiseon Kim

Insoo Koo

Library of Congress Cataloguing-in-Publication Data
A catalog record for this book is available from the Library of Congress.

British Library Cataloguing in Publication Data
Kim, Kiseon
CDMA systems capacity engineering—(Artech House mobile communications series)
1. Code division multiple access
I. Title II. Koo, I. S.
621.3'8456

ISBN 1-58053-812-6

Cover design by Yekaterina Ratner

685 Canton Street
Norwood, MA 02062

International Standard Book Number: 1-58053-812-6

10 9 8 7 6 5 4 3 2 1

Contents

Preface

Technology must be sustainable in the sense of efficiency, not only to satisfy quality requirements, but to obtain the same objectives with the minimum resources. Quality satisfaction has been an interesting issue to engineers as an objective of target technology, and technologies are continually evolving to optimize and fulfill the required qualities. The satisfaction objectives of quality can be quantitatively modeled in many cases. There had been continuous improvement of the satisfaction level on the modeled spaces, because the modeled problem is rather concrete and resolvable analytically within the artificially configured world. However, the sustainability relevant to the minimum resources is suggested by a higher layer than typical engineering, and it is rather an abstract topic for social movement and ecopolitical campaigns. Subsequently, while the engineers devote their time and efforts in the narrow concept of quality optimization, there have been growing concerns about whether the engineering development and relevant results are really contributive sustainably for mundane usages or simply for the progressing toward endless goals. Observing that global resources are becoming more scarce, it would be greatly beneficial if engineers really understand the issues of sustainability to implement technologies and systems.

Communications is an indispensable technology to process and transmit information. Obviously, communication technology needs to be sustainable in the sense of efficiency, not only to preserve the information within the quality requirements, but also to express the same contents with the minimum resources. Observing that the global resources of communication technology, such as frequencies and energy, are diminishing further and further, it will be greatly beneficial if engineers really understand the issues of sustainability to implement communication systems and satisfactory system performance. The communication resources can be represented by virtue of *capacity*, and quantitative expressions of capacity can be implemented by such sentences as:

- How many users can be included in a communication system as an indication of the capacity of the system?
- How many calls can be handled by a communication system as an indication of the capacity of the system?

By pondering the capacity issues of communication systems, along with various quality requirements such as transmission error rate, transmission speed, necessary bandwidth, and required power, we may develop sustainable systems, optimized mundane technologies beneficially both for technology consumers and for producers.

The code division multiple access (CDMA) communication system is a well-established technology in the sense that it is one of technically proven methods to transmit voice information for multiple users via wireless communications during the last decade. Further, CDMA is an emerging technology for next generation multimedia information of real-time and nonreal-time traffic and various multisource multitraffic communications environments. We have envisioned that CDMA is a key technology to satisfy the mundane usage of information transmission, and we are devoted to refining the definitions of *capacity* of the CDMA systems as the proper analytic measure to optimize the resources. At first, we need to observe the behavior of the voice and multimedia traffic to relate the simple measure of capacity and the characterizing parameters of traffic, where we specifically concentrated on the *traffic activity* and *activity factor* of the traffic. Also, *sensitivity*, a key issue in system engineering, is reinterpreted for the system capacity of the CDMA system to understand the nonideal parametric environment of system design. Once the capacity represents the objective for the system resource, while activity is the key parameter to represent traffic, the well-known capacity formula of an IS-95-type voice-only CDMA system can be revisited by our language. Naturally, we can extend the known results to general cases, including:

1. Multiple traffic cases;
2. Imperfect power control environment;
3. Delay requirements;
4. Limited system hardware resources;
5. Systems with multiple sectors and multiple frequency allocation (FA).

The CDMA system capacity is limited by the call processing algorithm and resource management, which is further analytically investigated for practical applications into traffic engineering, along with emerging environments. We consider that a service may be provided efficiently under hybrid frequency division multiple access (FDMA)/CDMA systems and the overlaying multiaccess systems, respectively.

Acknowledgments

All of the fruitful results in this book were possible under the supportive CDMA team environment in Kwangju Institute of Science and Technology (K-JIST), where authors, Dr. Yang, Jeong Rok, and many other team members were really enjoying the beauty of CDMA technology. Although this small book is a research summary of our understanding about CDMA technology, we believe that this is a small promise that we are working on the resource sustainability for the mundane usage. We would like to cherish each other on our various efforts of collaboration and valuable discussions that resulted in this book, and we expect further results to enhance the mundane value of CDMA technology for anybody at any time. Also, there was consistent support from various industry partners—SKT, Samsung, ADD, ETRI, IITA, MIC, and MOST, to name a few, without which it would not be possible to show this book to the CDMA technology world.

Last, but not least, we would like to thank all of the families of our CDMA team members for their silent understanding and endless support of what we have been doing, when we were not able to share any family life with them at all and have shown inconceivable behaviors for last several years to produce this work.

CHAPTER 1
Introduction

Since the telephone was invented in the late nineteenth century, there has been a steady development of telephone services, and the number of subscribers has continuously increased. One of the most revolutionary developments in telephone service in the late twentieth century was the introduction of the cellular variety of mobile phone services. As the number of subscribers has explosively grown in the wireless communication systems, provision of the mobility in telephone service was made possible by the technique of wireless cellular communication. As the bandwidth over the wireless link is a scarce resource, one of the essential functions of wireless communication systems is multiple access technique for a large number of users to share the resource.

Conceptually, there are mainly three conventional multiple access techniques: FDMA, time division multiple access (TDMA), and CDMA, as illustrated in Figure 1.1. The multiple access technique implemented in a practical wireless communication system is one of the main distinguishing characteristics of the system, as it determines how the common transmission medium is shared among users. FDMA divides a given frequency band into many frequency channels and assigns a separate frequency channel on demand to each user. It has been used for analog wireless communication systems. The representative FDMA wireless cellular standards include Advanced Mobile Phone System (AMPS) in the United States, Nordic Mobile Telephones (NMT) in Europe, and Total Access Communications System (TACS) in the United Kingdom [1]. TDMA is another multiple access technique employed in the digital wireless communication systems. It divides the frequency band into time slots, and only one user is allowed to either transmit or receive the information data in each slot. That is, the channelization of users in the same frequency band is obtained through separation in time. The major TDMA standards contain Global System Mobile (GSM) in Europe and Interim Standard 54/136 (IS-54/136) in North America [2]. GSM was developed in 1990 for second generation (2G) digital cellular mobile communications in Europe. Systems based on this standard were first deployed in 18 European countries in 1991. By the end of 1993, it was adopted in nine more European countries, as well as Australia, Hong Kong, much of Asia, South America, and now the United States.

CDMA is another multiple access technique utilized in the digital mobile communication systems. In CDMA, multiple access is achieved by assigning each user a pseudo-random code (also called pseudo-noise codes due to noise-like autocorrelation properties) with good auto- and cross-correlation properties. This code is used to transform a user's signal into a wideband spread spectrum signal. A receiver then transforms this wideband signal into the original signal bandwidth using the same

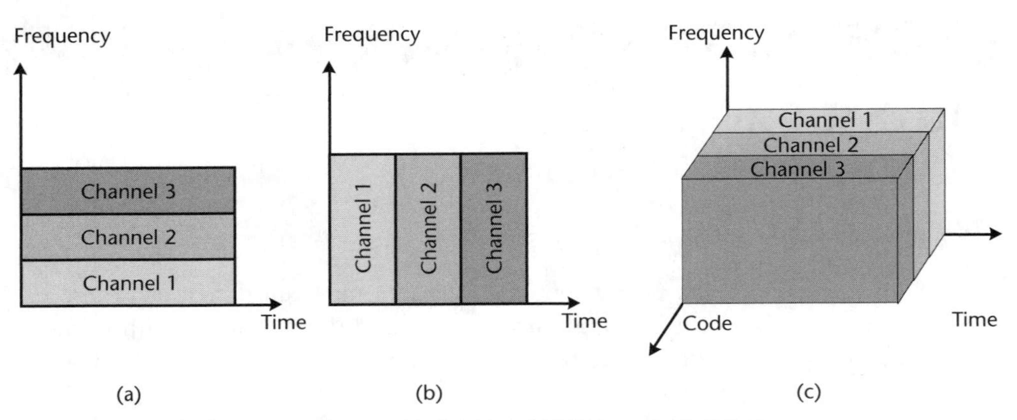

Figure 1.1 Multiple access schemes: (a) FDMA, (b) TDMA, and (c) CDMA.

pseudo-random code. The wideband signals of other users remain wideband signals. Possible narrowband interference is also suppressed in this process. The available spectrum is divided into a number of channels, each with a much higher bandwidth than the TDMA systems. However, the same carrier can now be used in all cells, such that the unity resource factor can be achieved in CDMA systems. It assigns each user a unique code, which is a pseudo-random sequence, for multiple users to transmit their information data on the same frequency band simultaneously. The signals are separated at the receiver by using a correlator that detects only signal energy from the desired user. One of the major CDMA standards is IS-95 in North America [3]. The use of CDMA technology in wireless cellular systems began with the development of the IS-95 standard [3], one of the 2G systems, in the early 1990s. At that time, the focus was to provide an efficient alternative to systems based on the AMPS standard in providing voice services, and only a low bit rate of 9.6 Kbps was provided. The main markets of IS-95 are the United States, Japan, and Korea, the latter being the largest market, with over 25 million subscribers. The success of IS-95 in Korea is based on the adoption of IS-95 as a national standard in the early 1990s. Now, CDMA is considered as one of the fastest growing digital wireless technologies. CDMA has been adopted by almost 50 countries around the world. Furthermore, CDMA was selected as a multiple-access scheme for the third generation (3G) system [4–6].

In addition to FDMA, TDMA, and CDMA, orthogonal frequency division multiplexing (OFDM), a special form of multicarrier modulation, can be used for multiplexing for multiple users. In OFDM, densely spaced subcarriers with overlapping spectra are generated using fast Fourier transform (FFT), and signal waveforms are selected in such a way that the subcarriers maintain their orthogonality despite the spectral overlap. One way of applying OFDM to the multiple access is through OFDM-TDMA or OFDM-CDMA, where different users are allocated different time slots or different frequency spreading codes. However, each user has to transmit its signal over the entire spectrum. This leads to an averaged-down effect in the presence of deep fading and narrowband interference. Alternatively, one can divide the total bandwidth into traffic channels (one or a cluster of OFDM subcarriers) so that multiple access can be accommodated in a form of the combination of OFDM and FDMA, which is called orthogonal frequency division multiple access (OFDMA).

An OFDMA system is defined as one in which each user occupies a subset of subcarriers, and each carrier is assigned exclusively to only one user at any time. Advantages of OFDMA over OFDM-TDMA and OFDM-CDMA include elimination of intracell interference and exploitation of network/multiuser diversity.

Space division multiple access (SDMA) is also recognized as a promising multiple access technology for improving capacity by the spatial filtering capability of adaptive antennas. SDMA separates the users spatially, typically using beamforming techniques such that in-cell users are allowed to share the same traffic channel. SDMA is not an isolated multiple access technique, but it can be applied to all other multiple access schemes [7]. In other words, a system that provides access by dividing its users in frequency bands, time slots, codes, or any combination of them, can also reuse its resources by identifying the user's positions so that under a given criterion, they can be separated in space.

CDMA techniques offer several advantages over other multiple access techniques, such as high spectral reuse efficiency, exploitation of multipath fading through RAKE combining, soft handoff, capacity improvements by the use of cell sectorization, and flexibility for multirate services [8–10]. The use of the CDMA techniques in wireless cellular communications commenced with the development of the IS-95A standard [3], of which IS-95A has been designed to achieve higher capacity than the first generation (1G) systems in order to accommodate rapidly growing subscribers. Further development of IS-95A toward higher bit rate services was started in 1996. This led to the completion of the IS-95B standard in 1998. While the IS-95A standard uses only one spreading code per traffic channel, IS-95B can concatenate up to eight codes for the transmission of higher bit rates. IS-95B systems can support medium user data rates of up to 115.2 Kbps by code aggregation without changing the physical layer of IS-95A. The next evolution of CDMA systems has led to wideband CDMA.

Wideband CDMA has a bandwidth of 5 MHz or more. Several wideband CDMA proposals have been made for 3G wireless systems. The two wideband CDMA schemes for 3G are WCDMA, which is network asynchronous, and cdma2000, which is synchronous. In network asynchronous schemes, the base stations (BSs) are not synchronized; in network synchronous schemes, the BSs are synchronized to each other within a few microseconds. Similar to IS-95, the spreading codes of cdma2000 are generated using different phase shifts of the same M sequence. This is possible because of the synchronous network operation. Because WCDMA has an asynchronous network, different long codes rather than different phase shifts of the same code are used for the cell and user separation. The code structure further impacts how code synchronization, cell acquisition, and handover synchronization are performed. The race of the high-speed packet data in CDMA started roughly in late 1999. Before then, WCDMA and cdma2000 systems supported packet data, but the design philosophy was still old in the sense that system resources such as power, code, and data rate were optimized to voice-like applications [11]. There has been a change since late 1999, as system designers realized that the main wireless data applications will be Internet protocol (IP)–related; thus, optimum packet data performance is the primary goal for the system designers to accomplish. With the design philosophy change, some new technologies have appeared, such as 1x radio transmission technology evolution for high-speed data only (1xEV-DO) and high-speed downlink packet access (HSDPA). Key concepts of these systems include adaptive and variable rate transmission, adaptive modulation and coding, and hybrid automatic repeat request (ARQ) to adapt the IP-based network for a given channel condition and workload with the objective of maximizing the system performance by using various adaptive techniques while satisfying the quality of service (QoS) constraints. First, HSDPA is a major evolution of WCDMA wireless network, where the peak data rate and throughput of the WCDMA downlink for best effort data is greatly enhanced when compared to release 99.

In March 2000, a feasibility study on HSDPA was approved by 3GPP. The study report was part of release 4, and the specification phase of HSDPA was completed in release 5 at the end of 2001. By contrast, cdma2000 is followed by 1xEV-DO for the first phase, in the sense of deployment schedule, and high-bit-rate data and voice (1xEV-DV) for the second phase. It is noteworthy that 1xEV-DVdoes not necessarily follow 1xEV-DO. Both 1xEV-DO and 1xEV-DV allow data rates of up to 2.4 Mbps in 1.25-MHz bandwidth, compatible with the frequency plan of 2G and 3G CDMA systems based on IS-95 and cdma2000. Figure 1.2 illustrates the evolution of 2G/3G cellular and the revolutionary step toward future wireless systems.

It is not hard to see the reasons for the success of CDMA. Its advances over other multiple-access schemes include higher spectral reuse efficiency due to the unity reuse factor, greater immunity to multipath fading, gradual overload capability, and simple exploitation of sectorization and voice inactivity. Moreover, CDMA has more robust handoff procedures [12–15].

Because wireless systems have limited system resources and multimedia services have various QoS requirements, the evaluation of the network system capacity is one of important issues for facilitating multimedia communications among multiple users. The capacity of CDMA systems is closely related to traffic characteristics, power control, sectorization, and other factors. It is an interesting topic to evaluate the capacity of CDMA systems supporting mixed services, focusing on the characteristics of various kinds of traffic. In this book, we tackle this issue especially for IS-95-like and cdma2000-like CDMA systems where the BSs are all synchronized. All contents in the book, however, can be applied to WCDMA-like systems that have an

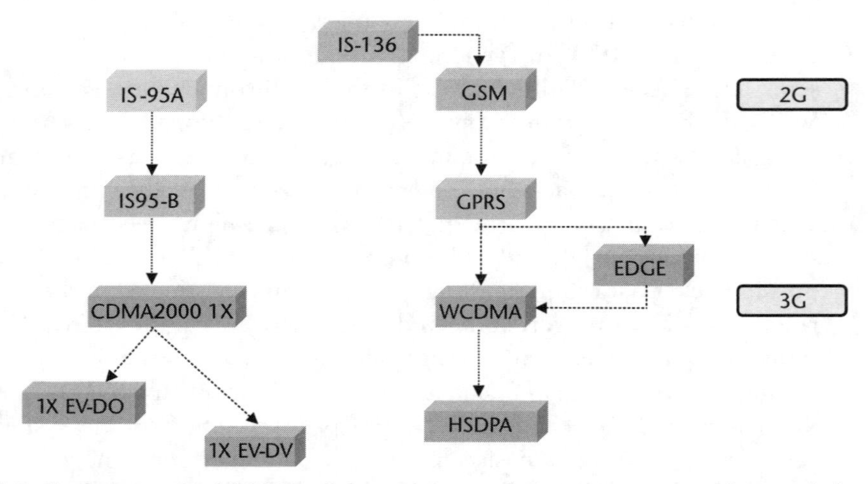

Figure 1.2 Evolution path of 2G/3G cellular and the revolutionary step toward future wireless systems.

asynchronous network if the asynchronous aspects such as code synchronization, cell acquisition, and handover synchronization are properly considered when evaluating the capacity.

Before we deal with CDMA capacity issues in more detail, let's consider some basic elements of CDMA systems. Figure 1.3 shows the basic elements required to process a call in the CDMA system, including the mobile switching center (MSC), the BS controller (BSC), and mobile stations (MS). Their proper combination is essential for the efficient deployment of a CDMA system toward a tradeoff in the cost of each subsystem and its scalability for future expansion.

The MSC is the core of the CDMA systems, the main functions of which include switching functions between mobile calls; switching calls between a mobile and the outside networks, such as the public switched telephone network (PSTN), public data network (PDN), or integrated service digital network (ISDN); as well as network maintenance, such as MS user location registration, MS equipment registration, authentication, and roaming. The BSC includes all of the radio transmission and reception equipment, namely base transceiver subsystems (BTS), to handle a wireless call from the MS according to the given wireless protocol within the proper cell range, and the control functions of cell configuration, handover, power control, and supervision of multiple BTSs. Under the wireless protocol, each call signal is processed on the channel element (CE) in the BTS, the processing of which can be classified into two phases: chip-rate processing and symbol rate processing.

On the CE, there is a complex mix of the dataflow and control processing, and as a call proceeds from the antenna towards the backhaul of the system, the control processing has more significance than the dataflow processing in the sense of resource utilization. Typically, the dataflow processing of a call is very hardware intensive and is well suited to dedicated programmable hardware solutions, while the call processing is better suited for implementation using either hardware state machines or software on a control processor. While the mobile communications

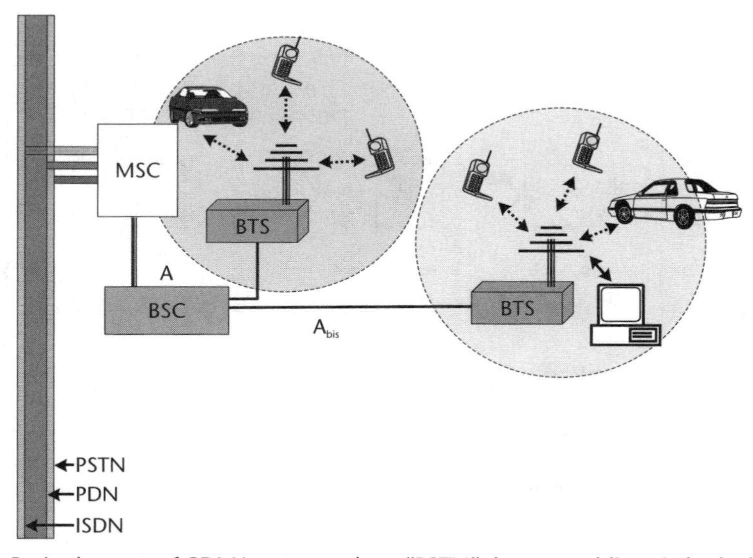

Figure 1.3 Basic elements of CDMA systems where "PSTN" denotes public switched telephone network, "PDN" denotes public data network, and "ISDN" denotes integrated service digital network. "A" and "A_{bis}" are the interface between MSC and BSC and between BSC and BTS, respectively.

evolve, the channel card in the BTS—which includes a set of channel elements—needs to be flexible to address the flexibility requirements driven by the diverse standards and various communication signal-processing techniques, such as multiuser detection (MUD) and beamforming. For example, MUD, also called joint detection and interference cancellation, provides means of reducing the effect of multiple access interference where all signals would be detected jointly or interference from other signals would be removed by subtracting them from the desired signal such that MUD increases the system capacity. The capacity of CDMA systems is related to the interference level such that adopting SDMA in the CDMA systems will produce an overall performance enhancement. In certain SDMA, beamforming technologies are adopted to implement smart antennas. Smart antennas are multibeam or adaptive array antennas without handover between beams. Multibeam antennas use multiple fixed beams in a sector, while in an adaptive array the received signals by the multiple antennas are weighted and combined to maximize the signal-to-noise ratio (SNR). A multibeam antenna with M beams can increase the capacity by a factor of M by reducing the number of interferences, while adaptive arrays can provide some additional gain by suppressing interferes further.

Implementations would be based on field-programmable gate arrays (FPGAs) for the dataflow processing and programmable digital signal processors (DSPs) for the control processing, while application-specific integrated circuits (ASICs) are an attempt to reduce costs. Thus, all the chip-rate processing and some symbol-rate processing in the CE card resides on the FPGA, and the rest of the symbol-rate processing and some layer 1 control resides on the DSP, as shown in Figure 1.4.

1.1 Capacity Issues

The capacity of CDMA systems is an extremely important issue in terms of its economic viability because the overall revenue of the operator is proportional to the

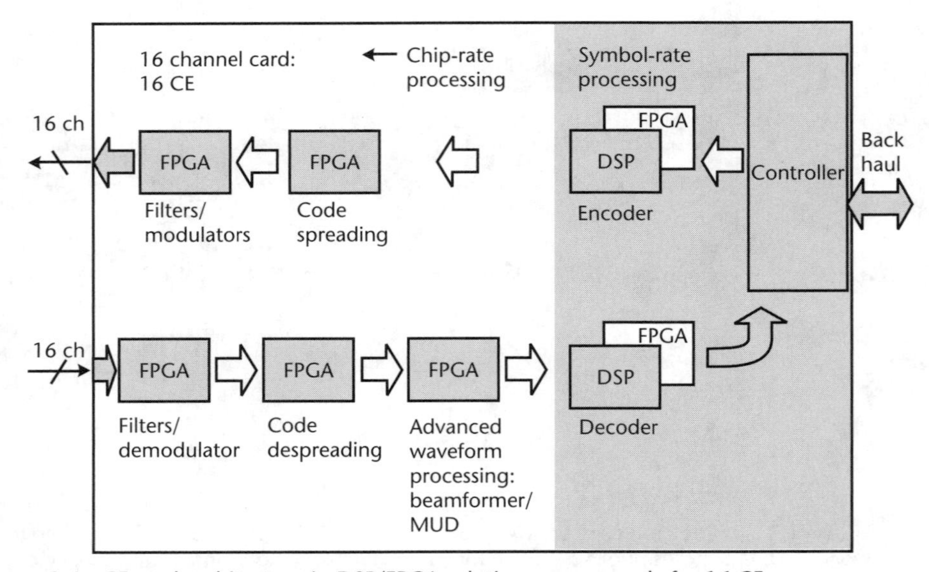

Figure 1.4 CE card architecture in DSP/FPGA solution—an example for 16 CEs.

system capacity. For example, in the simplest case, where all users are provided with the same service offering for the same cost, the revenue of the operator will be maximized if the operator maximizes the number of users in the system, even though the revenue certainly depends on economic factors such as the price and competing operators or services and on the technical limitation of the systems [16, 17]. Another useful application of the system capacity is the system dimensioning. For example, when capacity is evaluated as a function of various system parameters, we may dimension the required size of the target system parameters to accommodate the target offered traffic load.

The capacity of a CDMA system can be defined in several ways. One of its typical definitions is the maximum number of simultaneous users that can be supported by the system while the service quality requirements of each user, such as the data rate, bit error rate (BER), and outage probability, are being satisfied. In the case of FDMA or TDMA systems, the number of frequency slots or the number of time slots corresponds to the system capacity, respectively, as TDMA and FDMA systems tend to run out of frequency channels or time slots before they become capacity or coverage limited. On the other hand, in the case of CDMA systems, transmit power constraints and the system's self-generated interference ultimately restrict CDMA capacity, as CDMA systems tend to be capacity or coverage limited before they run out of codes and such. For example, the reverse link reaches capacity when a mobile station has insufficient transmit power to overcome the interference from all other mobile stations to meet the required ratio of bit energy to interference power density at the intended BS. Similarly, in the forward link, capacity is reached when the total power required to successfully transmit to all mobile stations hosted by the cell exceeds BS power in order to meet the required ratio of bit energy to interference density at all intended mobile stations.

Lots of research exists to find the maximum number of simultaneous users that CDMA systems can support while maintaining desired QoS. The capacity of voice-only CDMA systems can be found [18]. In [19], V. K. Paulrajan et al. investigated the capacity of CDMA systems for multiclass services in single cell case and visualized the resulting capacity. Further, J. Yang et al. expanded the approach of [19] to the case of multicells [20].

The capacity of CDMA systems with respect to the possible number of supportable users can be utilized for radio resource management, such as call admission control (CAC) or resource allocation for ongoing calls as well as for a measure of revenue of the operator. For example, when a new user requests a service, the system resource required by the user can be expected. If the system resource required by the user is smaller than the remaining system resources, then the user is accepted. Otherwise, it will be blocked. In such a case, the evaluated system capacity bounds can be used as a reference for the threshold of CAC. Furthermore, the capacity bound can be used for system resource management. If current users in the system do not use all of the system resources, the remaining system resources may be allocated to the current users to increase system throughput or quality until a new user requests a service and the system allocation is newly configured to accept the user.

For the purpose of controlling the system, rather than estimating the supportable size of the system, alternatively the capacity measure is the average traffic load that can be supported with a given quality and with availability of service as measured by the blocking probability. The average traffic load in terms of the average number of users requesting service and further resulting in the target blocking probability is called as the *Erlang capacity*. Regarding the evaluation of Erlang capacity, Viterbi and Viterbi reported the Erlang capacity of CDMA systems only for voice, based on outage probability where the outage probability is defined as the probability that the interference plus noise power density I_o exceeds the noise power density N_o by a factor $1/\eta$, where η takes on typical values between 0.25 and 0.1 [21]. In [22], Sampath et al. extended the results of Viterbi to CDMA systems supporting voice and data calls.

Viterbi's model for Erlang capacity is a $M/M/\infty$ queue with voice activity factor, $\rho(\rho = 0.4)$ (i.e., a queue model with Poisson input and with infinite service channels that are independent and identically distributed. Exponential service time distribution is considered, where M and M means that each user has exponentially distributed interarrival times and service times, and ∞ means infinite number of available servers. More fundamental explanations on $M/M/\infty$ queue are available in Appendix A. Because the capacity of a CDMA system is soft, Viterbi and Viterbi prefer outage probability to blocking probability. The resulting expression for outage probability is simply the tail of the Poisson distribution.

$$P_{out} < e^{-\frac{\rho\lambda}{\mu}} \sum_{k=K'_0}^{\infty} \left(\frac{\rho\lambda}{\mu}\right)^k / k! \tag{1.1}$$

where K'_0 satisfies the outage condition

$$\sum_{j=2}^{m} v_j < \frac{W / R(1-\eta)}{E_b / I_o} = K'_0 \tag{1.2}$$

and v_j is the binary random variable indicating whether the jth voice user is active at any instant. For example, for a process gain of 128, $\eta = 0.1$, and $E_b/N_0 = 5$, $K'_0 = 23$. If voice activity factor is 1, the maximum number of users supported is $m = K'_0 + 1 = 24$.

Viterbi and Viterbi basically presumed outage probability to call blocking probability. However, the outage probability does not directly correspond to the call blocking, as call blocking is mainly caused when a call is controlled by a CAC rule. That is, blocking and outage should be distinguished when evaluating the Erlang capacity because blocking occurs when an incoming mobile cannot be admitted in the system, while outage occurs when a mobile admitted in the cell cannot maintain the target QoS requirement.

One approximate method to evaluate the Erlang capacity of CDMA systems is to use an $M/M/m$ loss model [23–25] (i.e., m server model with Poisson input and exponential service time such that when all of the m channels are busy, an arrival leaves the system without waiting for service), where M and M means that each user has exponentially distributed interarrival times and service times, and m means there is m finite number of available servers. More fundamental explanations on $M/M/m$ queue are available in Appendix B. The blocking probability of the $M/M/m$ model is simply given by the Erlang B formula, rather than the Poisson distribution, but the Poisson distribution and Erlang B formula practically arrive at the same results when number of servers in the system is larger than 20 [23]. Unlike the approach of [21], this approach allows for the provision of different grades of service for different types of calls. This is made possible by the introduction of a new grade of service metric, the blocking probability in addition to the outage probability [25].

This Erlang analysis of the CDMA systems can be performed in two stages. In the first stage, we determine the number of available servers, or available *virtual* trunk channels. In the second stage, we calculate the Erlang capacity from the number of virtual trunk channels. The trunk channels are not physical trunk channels but rather virtual ones. Noting that the limitation of the underlying physical system is taken into account when evaluating the number of available trunk channels, we can refer to the trunking capacity as the maximum possible number of simultaneous users that can be supported by the system while the QoS requirements of each user (e.g., data rate, BER, and outage probability) are being satisfied.

This approximate analysis method is simpler when calculating the Erlang capacity of CDMA systems than Viterbi's one due to the following reasons:

- *First stage.* As a trunk capacity, we can utilize the capacity analysis results regarding the possible number of simultaneous users that can be handled in the system for given QoS requirements, such as data rate, BER and target outage probability, which have been researched in many other papers [18–20].
- *Second stage.* When determining the Erlang capacity from the number of virtual trunk channels, we can utilize the loss network model and its results, which are already well developed in the circuit-switched network.

Another alternative definition of the system capacity is the sum of throughput and the Erlang capacity [26]. This measure is particularly useful when the data users have best effort applications and further share the network resources with real-time traffic like voice. Best effort applications such as file transfer and electronic mail can adapt their instantaneous transmission rate to the available network resources and thus need not be subject to admission control. On the other hand, real-time applications need some guaranteed minimum rate as well as delay bounds, which require reservation of system capacity such that real-time traffic is subject to CAC. In [26], Sato et al. analyzed the capacity of an integrated voice and data system over a CDMA unslotted ALOHA with channel load sensing protocol (CLSP) and investigated the effect of the threshold for the number of data transmissions on the capacity of CDMA unslotted ALOHA systems.

1.2 Overview and Coverage

The commercial CDMA systems are mainly classified into two groups. One group is the synchronized CDMA systems, such as IS-95-like and cdma2000-like systems. The other group is the unsynchronized CDMA systems, such as WCDMA-like systems.

In this book, we are mainly concerned with evaluating the capacity of the synchronized CDMA systems in various aspects of capacity definition. All contents in the book, however, can be applicable to WCDMA-like systems that have an asynchronous network if the asynchronous aspects such as code synchronization, cell acquisition, and handover synchronization are properly considered when evaluating the capacity. The remaining part of this book consists of 11 chapters. In this section, we present the organization of this book and, outline the important contributions of each chapter.

In Chapter 2, the capacity of CDMA systems supporting various service classes is analyzed with respect to the maximum number of simultaneous users where each user is characterized by its own QoS requirements. In the multiclass CDMA systems, the QoS requirements are composed of a quality (E_b/N_0) requirement and a transmission rate requirement [27, 28]. Different services require different received signal power levels; thus, the amount of interference generated by one service user is different from that generated by another service user. The upper limit for the number of users of one service subsequently is limited by the numbers of users in the other services. To fully utilize the multimedia CDMA system resources, the system capacity must be identified, and correct tradeoffs are required between the number of users in each service. In this chapter, we tackle analyzing the capacity of a CDMA system supporting multiclass services such that a simple upper-limit hyperplane concept is formulated to visualize the capacity of a multimedia CDMA system. Further, the tradeoffs between the level of system resources needed for a certain user and that needed for others are illustrated analytically within the concept of resource management. The results of this chapter will be utilized in remaining chapters of this book to evaluate the Erlang capacity and propose the resource management schemes of CDMA systems.

In Chapter 3, sensitivity analysis of capacity parameters on CDMA system capacity is presented. CDMA system capacity can be expressed as a function of various parameters such as required E_b/N_0, traffic activity factor, processing gain, system reliability, frequency reuse factor, and power control error. The sensitivity of respective parameters on the CDMA system capacity can afford a proper tool to design CAC scheme, particularly when the capacity limit is utilized for a reference to threshold for CAC schemes. In this chapter, we adopt the sensitivity analysis methodology and present the sensitivity of the system capacity with respect to the system reliability, as an example of sensitivity analysis in CDMA systems such that the effects of the system reliability as well as the imperfection due to the imperfect power control on the reverse link system capacity of multimedia CDMA systems are evaluated in explicit way.

In Chapter 4, the effect of traffic activity on the system capacity is analyzed. As the capacity of a CDMA system is interference limited, any reduction of the interference improves the system capacity [18]. One of the techniques to reduce the interference is to operate the system in a discontinuous transmission mode (DTX) for the traffic with ON/OFF traffic activity [29]. In the DTX mode, the transmission can be suppressed when there is no data to be sent (i.e., the user is in an idle, or OFF, state, which causes the interference to be reduced). The simplest way to include this reduction of the interference due to the traffic activity in the capacity analysis is to consider the long-term average interference, in which the random characteristics of traffic activity are assumed to be simplified to the mean of traffic activity, (i.e., the traffic activity factor). For instance, the interference was assumed to be averaged out and reduced by a factor of the reciprocal of the voice traffic activity factor for a preliminary capacity analysis for a voice-only CDMA system [18]. In Chapter 2, the same assumption was used to analyze the capacity of a voice/data CDMA system. However, because the probability that the interference is above the average interference is not negligible, a more practical way is to statistically consider the fluctuation of the interference due to the traffic activity by modeling the traffic activity as a binomial random variable [18, 25]. In this chapter, we subsequently compare the capacity analyzed with the latter way with that analyzed with the former way. We further investigate the overall dependency of the system capacity on the traffic activity under the same transmission rate and under the same average rate. According to the activity factor, the average rate and the transmission rate are changed under the same transmission rate and under the same average rate, respectively.

With the growing demands for multimedia services and the high degree of user mobility, radio resource management (RRM) plays an important role in CDMA systems to efficiently utilize the limited radio resources and to provide more mobile users with guaranteed QoS. Major RRM schemes can be divided into CAC and resource allocation for ongoing calls [17, 30, 31]. CAC involves control of both new calls and handoff calls, and the resource allocation for ongoing calls is to distribute the radio resources among existing users so that the system objective functions, such as the throughput, can be maximized while maintaining the target QoS. This book also addresses the RRM in CDMA systems supporting multiclass services from these two perspectives. First, Chapter 5 proposes a resource allocation scheme with which we can find the optimum set of data rates for concurrent users and further maximize the system throughput while satisfying the minimum QoS requirements of each user for ongoing connected calls. Second, Chapter 6 presents a CAC scheme for CDMA systems supporting voice and data services to accommodate more traffic load in the system, where some system resources are reserved exclusively for handoff calls to have higher priority over new calls, and queuing is allowed for both new and handoff data traffic that is not sensitive to delay. More details on Chapters 5 and 6 are as follows: In Chapter 5, an efficient resource allocation scheme is proposed to efficiently utilize the remaining system resources. In most cases, the system is not being situated on the capacity limit in terms of the number of concurrent users, and thus there exist some remaining resources. For the efficient use of the system capacity, the system could be designed to allocate the remaining system resources. As the capacity of a CDMA system is interference limited, the remaining system resources can be interpreted as power (E_b/I_0) or data rate. For dual-service classes composed of a constant bit rate (CBR) service class and a variable bit rate (VBR) service class, a resource allocation scheme has been proposed to maximize the throughput by allocating the remaining system resources to the limited number of users rather than all users in the VBR service class [32]. In this chapter, for CDMA systems supporting multiclass services, the relationship between the data rates of VBR service classes is investigated under the condition that all users' QoS requirements are satisfied, and a simple scheme optimally allocating the remaining system resources by selecting a VBR class is presented to maximize the throughput. We further observe to which group the remaining system resources should be allocated so as to maximize the throughput, according to the parameters of the VBR service class, such as the number of users and the QoS requirements.

In Chapter 6, we propose a CAC scheme for the CDMA systems supporting voice and data services taking into account user mobility and traffic characteristics. Moreover, we analyze the Erlang capacity under the proposed CAC scheme. In the proposed CAC scheme, some system resources are reserved exclusively for handoff calls to have higher priority over new calls. Additionally, queuing is allowed for both new and handoff data traffic that is not sensitive to delay. The proposed CAC scheme is based on the idea of reservation and queuing, and there are many relevant papers [33–37]. Particularly, the scheme in [37] seems to be very similar to the proposed scheme. However, noting that [37] considered the buffer for handoff voice calls, and that voice traffic is delay sensitive, it is not efficient to utilize the buffer for handoff voice calls. In the proposed scheme, we consider the buffer for new data calls rather than voice calls, as the data traffic is more tolerable to the delay requirement. Furthermore, the Erlang capacity of CDMA under the proposed CAC is evaluated, and the procedure for properly selecting the CAC-related parameters, such as the number of reservation channels and queue lengths, is presented.

In FDMA and TDMA systems, traffic channels are allocated to calls as long as they are available. Incoming calls are blocked when all channels have been assigned. The physical parallel in CDMA systems is for a call to arrive and find that the BS has no receiver processors left to serve it. However, often a more stringent limit on the number of simultaneous calls is determined by the total interference created by the admitted users exceeding a threshold. Outage in CDMA systems is said to occur when the interference level reaches a predetermined value above the background noise level. In a CDMA system, a CE performs the baseband spread spectrum signal processing of a received signal for a given channel (pilot, sync, paging, or traffic channel). Practically, CDMA systems are equipped with a finite number of CEs, which is afforded by cost-efficient strategies, as the CE is a cost part of the BS, which introduces inherently hard blocking in CDMA systems.

Subsequently, Erlang capacity is determined not only by the maximum number of simultaneous active users but also by the maximum number of CEs available for traffic channels. In this book, we analyze the Erlang capacity of CDMA systems with the consideration of the limited number of CEs in BSs as well as without the limitation on the CEs in BSs. First, Chapter 7 tackles the Erlang capacity of CDMA systems supporting multiclass services for the case of no limitation of the CEs in BSs, based on a multidimension $M/M/m$ loss model. For an IS-95-type CDMA system supporting voice/data services, the Erlang capacity limits are depicted in conjunction with a two-dimensional Markov chain. Further, the channel reservation scheme is considered to increase total Erlang capacity by balancing the Erlang capacities with respect to voice and data services. Chapter 8 is also devoted to evaluating the capacity of CDMA systems supporting voice and data services under the delay constraint. To achieve higher capacity using the delay-tolerant characteristic, data calls can be queued until the required resources are available. The blocking probability and the average delay have been typically considered performance parameters for the delay-tolerant traffic [38, 39]. In Chapter 8, we introduce a new performance measure, the delay confidence, as the probability that a new data call is accepted within the maximum tolerable delay without being blocked. The Erlang capacity is defined as a set of average offered loads of voice and data traffic that can be supported while the required blocking probability for voice traffic and the required delay confidence for data traffic are satisfied. To analyze the Erlang capacity with the first-come first-served service discipline, a two-dimensional Markov model is used where the waiting is allowed in the queue with a finite size for the data calls. Based on the Markov model, we develop the procedure to analyze the delay confidence of data calls.

After that, the remaining chapters deal with the capacity evaluation of CDMA systems with consideration to both the limitation on the maximum number of CEs available in BS and the limitation on the maximum number of simultaneous active users in the air link. More specifically, Chapter 9 presents the effect of the limited number of CEs in BSs on the Erlang capacity of CDMA systems supporting multi-class services as an expansion of Chapters 7 and 8. In addition, a graphic interpretation method will also be presented for the multiple FAs case, where the required calculation complexity of the exact method is too high to calculate the Erlang capacity of CDMA systems with high FAs. Chapter 10 presents an approximated method to calculate the Erlang capacity of CDMA systems with multiple sectors and multiple frequency allocation bands, in order to overcome the complexity problem of the exact calculation method proposed in the previous chapter. The proposed approximate analysis method reduces the exponential complexity of the old method [40] down to linear complexity for calculating the call blocking probability, and the results calculated by the proposed approximate method provide a difference only a few percent from the exact values, which makes the proposed method practically useful.

Future CDMA networks will combine with different radio access technologies such as WCDMA/UMTS, WiFi (IEEE 802.11), WiMax (IEEE 802.16), and even IEEE 802.20, and further will evolve into the multiaccess systems where several distinct radio access technologies coexist, and each radio access technology is called a *subsystem*. In multiservice scenarios, the overall capacity of multiaccess networks depends on how users of different services are assigned on to subsystems, as each subsystem has distinct features from each other with respect to capacity. For example, IS-95A can handle voice service more efficiently than data service, while 1xEV-DO can handle data service more efficiently than voice service.

In this book, we also tackle the Erlang capacity evaluation of multiaccess systems in two cases. First, in Chapter 11, we consider the case that each subsystem provides similar air link capacity. As a typical example, we consider hybrid FDMA/CDMA, where like FDMA the available wideband spectrum of the hybrid FDMA/CDMA is divided into a number of distinct bands. Each connection is allocated to a single band such that each band facilitates a separate *narrowband* CDMA system, whose signals employ direct sequence (DS) spreading and are transmitted in one and only one band. Subsequently, it can be assumed that each carrier will provide similar air link capacity. For evaluating the Erlang capacity for hybrid FDMA/CDMA systems, we consider two channel allocation schemes: independent carrier channel assignment (ICCA) scheme and combined carrier channel assignment (CCCA) scheme. In the ICCA scheme, traffic channels of each carrier are handled independently so that each MS is allocated a traffic channel of the same carrier as it used in its idle state. By contrast, the CCCA scheme combines all traffic channels in the system so that when a BS receives a new call request, the BS searches the least occupied carrier and allocates a traffic channel in that carrier. In [41], Song et al. analyzed and compared performances of the hybrid FDMA/CDMA system under ICCA and CCCA schemes. However, they focused only on the voice-oriented system and considered the call-blocking model in which the call blocking is caused only by a scarcity of CEs.

In this chapter, we consider the expanded blocking model, where call blocking is caused not only by a scarcity of CEs in the BS but also by insufficient available channels per sector. For each allocation scheme, the effect of the number of carriers of hybrid FDMA/CDMA systems supporting voice and data services on the Erlang capacity is observed, and the optimum values of the system parameters such as CEs are selected with respect to the Erlang capacity. Furthermore, the performances of ICCA are quantitatively compared with those of CCCA.

Second, in Chapter 12, we consider the case that each subsystem provides different air link capacity, as in the case with coexisting GSM/EDGE-like and WCDMA-like subsystems. In this case, the overall capacity of multiaccess networks depends on the employed service assignment (i.e., the way of assigning users of different services onto subsystems). In Chapter 12, two user assignment schemes are considered: the service-based assignment algorithm [42] as a best case reference, which roughly speaking assigns users to the subsystem where their service is most efficiently handled, and the rule opposite the service-based assignment as a worst case reference. These two cases will provide lower and upper limits of Erlang capacity of multiaccess systems under common operation method.

References

[1] Lee, W. C. Y., *Mobile Cellular Telecommunications*, New York: McGraw-Hill, 1995.

[2] Rappaport, T. S., *Wireless Communications,* Englewood Cliffs, NJ: Prentice-Hall, 2002.

[3] IS-95-A, "Mobile Station-Base Station Compatibility Standard for Dual-Mode Wideband Spread Spectrum Cellular System," 1995.

[4] Prasad, R., W. Mohr, andW. Konhauser, *Third Generation Mobile Communication Systems,* Norwood, MA: Artech House, 2000.

[5] IS-2000, "Physical Layer Standard for cdma2000 Spread Spectrum Systems," 2000.

[6] 3GPP, "Physical Channels and Mapping of Transport Channels onto Physical Channels (fdd)," *3G TS 25.211*, 1999.

[7] Ojanpera, T., and R. Prasad, *WCDMA: Towards IP Mobility and Mobile Internet,* Norwood, MA: Artech House, 2000.

[8] Prasad, R., *CDMA for Wireless Personal Communications,* Norwood, MA: Artech House, 1996.

[9] Adachi, F., M. Sawahashi, and H. Suda, "Wideband DS-CDMA for Next-Generation Mobile Communcations Systems," *IEEE Communications Magazine*, 1998, pp. 56–69.

[10] Dehghan, S., et al., "W-CDMA Capacity and Planning Issues," *Electronics & Communication Engineering Journal*, 2000, pp. 101–118.

[11] Qiu, R., W. Zhu, and Y. Zhang, "Third-Generation and Beyond (3.5g) Wireless Networks and its Applications," *IEEE Proc. of ISCAC*, 2002, pp. I-41–I-44.

[12] Hernando, J. M., and F. Perez-Fontan, *Introduction to Mobile Communications Engineering,* Norwood, MA: Artech House, 1999.

[13] Hammuda, H., *Cellular Mobile Radio Systems (Designing Systems for Capacity Optimization),* New York: John Wiley & Sons, 1997.

[14] Lee, W., "Overview of Cellular CDMA," *IEEE Trans. on Vehicular Technology*, 1991, pp. 291–302.

[15] Kohno, R., R. Meidan, and L. Milstein, "Spread Spectrum Access Methods for Wireless Communications," *IEEE Commun. Mag.*, 1995, pp. 58–67.

[16] Zander, J., "On the Cost Structure of Future Wideband Wireless Access" *IEEE Proc. of Vehicular Technology Conference*, 1997, pp. 1773–1776.

[17] Zander, J., and S. L. Kim, *Radio Resource Managment for Wireless Networks,* Norwood, MA, Artech House, 2001.

[18] Gilhousen, K. S., et al., "On the Capacity of a Cellular CDMA System," *IEEE Trans. on Vehicular Technology*, 1991, pp. 303–312.

[19] Paulrajan, V. K., J. A. Roberts, and D. L. Machamer, "Capacity of a CDMA Cellular System with Variable User Data Rates," *Proc. of IEEE Global Telecommunications Conference*, 1996, pp. 1458–1462.

[20] Yang, Y. R., et al., "Capacity Plane of CDMA System for Multimedia Traffic," *IEE Electronics Letters*, 1997, pp.1432–1433.

[21] Viterbi, A. M., and A. J. Viterbi, "Erlang Capacity of a Power-Controlled CDMA System," *IEEE Journal on Selected Areas in Communications*, 1993, pp. 892–900.

[22] Sampath, A., N. B. Mandayam, and J. M. Holtzman, "Erlang Capacity of a Power Controlled Integrated Voice and Data CDMA System," *IEEE Proc. of Vehicular Technology Conference*, 1997, pp. 1557–1561.

[23] Jacobsmeyer, J., "Congestion Relief on Power-Controlled CDMA Networks," *IEEE Journal on Selected Areas in Communications*, 1996, pp. 1758–1761.

[24] Koo, I., et al., "Analysis of Erlang Capacity for the Multimedia DS-CDMA Systems," *IEICE Trans. on Fundamentals*, 1999, pp. 849–855.

[25] Matragi, W., and S. Nanda, "Capacity Analysis of an Integrated Voice and Data CDMA System," *IEEE Proc. of Vehicular Technology Conference*, 1999, pp. 1658–1663.

[26] Sato, T., et al., "System Capacity of an Integrated Voice and Data CDMA Network in Channel Load Sensing Protocol," *IEEE Proc. of GLOBECOM*, 1997, pp. 899–903.

[27] Wu, J. S., and J. R. Lin., "Performance Analysis of Voice/Data Integrated CDMA System with Constraints," *IEICE Trans. on Communications*, Vol. E79-B, 1996, pp. 384–391.

[28] Sampath, A., P. S. Kumar, and J. M. Holtzman, "Power Control and Resource Management for a Multimedia CDMA Wireless System," *IEEE Proc. of International Symposium on Personal, Indoor and Mobile Radio Communications*, 1995, pp. 21–25.

[29] Timotijevic, T. and J. A. Schormans, "ATM-Level Performance Analysis on a DS-CDMA Satellite Link Using DTX," *IEE Proc.—Communications*, 2000, pp. 47–56.

[30] Tripathi, N. D., J. H. Reed, and H. F. VanLandingham, *Radio Resource Management in Cellular Systems*, Boston, MA: Kluwer Academic Publishers, 2001.

[31] Hong, D., and S. Rappaport, "Traffic Model and Performance Analysis for Cellular Mobile Radio Telephone Systems with Prioritized and Nonprioritized Handoff Procedures," *IEEE Trans. on Vehicular Technology*, 1986, pp. 77–92.

[32] Ramakrishna, S., and J. M. Holtzman, "A Scheme for Throughput Maximization in a Dual-Class CDMA System," *IEEE Journal on Selected Areas in Communications*, 1998, pp. 830–844.

[33] Hong, D., and S. Rappaport, "Traffic Model and Performance Analysis for Cellular Mobile Radio Telephone Systems with Prioritized and Nonprioritized Handoff Procedures," *IEEE Trans. on Vehicular Technology*, 1986, pp. 77–92.

[34] Del Re, E., et al., "Handover and Dynamic Channel Allocation Techniques in Mobile Cellular Networks," *IEEE Trans. on Vehicular Technology*, 1995, pp. 229–237.

[35] Hong, D., and S. Rappaport, "Priority Oriented Channel Access for Cellular Systems Serving Vehicular and Portable Radio Telephones," *IEE Proc. of Communications*, 1989, pp. 339–346.

[36] Pavlidou, F., "Two-Dimensional Traffic Models for Cellular Mobile Systems," *IEEE Trans. on Communications*, 1994, pp. 1505–1511.

[37] Calin, D., and D. Zeghlache, "Performance and Handoff Analysis of an Integrated Voice-Data Cellular System," *IEEE Proc. of PIMRC*, 1997, pp. 386–390.

[38] Koo, I., E. Kim, and K. Kim, "Erlang Capacity of Voice/Data DS-CDMA Systems with Prioritized Services," *IEICE Trans. on Communications*, 2001, pp. 716–726.

[39] Bae, B. S., K. T. Jin, and D. H. Cho, "Performance Analysis of an Integrated Voice/Data CDMA System with Dynamic Admission/Access Control," *IEEE Proc. of Vehicular Technology Conference*, Spring 2001, pp. 2440–2444.

[40] Kim, K. I., *Handbook of CDMA System Design, Engineering, and Optimization*, Upper Saddle River, NJ: Prentice Hall, 2000.

[41] Song, B., J. Kim, and S. Oh, "Performance Analysis of Channel Assignment Methods for Multiple Carrier CDMA Cellular Systems," *IEEE Proc. of VTC (Spring)*, 1999, pp. 10–14.

[42] Furuskar, A., "Allocation of Multiple Services in Multi-Access Wireless Systems," *IEEE Proc. of MWCN*, 2002, pp. 261–265.

CHAPTER 2

System Capacity of CDMA Systems

The maximum number of simultaneous users satisfying QoS requirements, a typical capacity definition in CDMA systems, should be evaluated in both single cell and multiple cell environments, as system capacity is a basic problem to research resource management and CAC. In this chapter, we tackle this issue in a CDMA system supporting multiclass services such that a simple upper-bounded hyperplane concept is formulated to visualize the capacity of a multimedia CDMA system. The tradeoffs between the level of system resources needed for a certain user and that needed for others are illustrated analytically within the concept of resource management.

2.1 Introduction

In recent years, communication systems for multimedia services such as voice, image, and data have been researched and developed in the wired communication system. The demand for multimedia services is expected to increase in the wireless communication system as well. The CDMA scheme has been proposed for a next generation wireless system that will offer multimedia services. In the wireless communication system, the system capacity, resource management, and CAC are to be considered for facilitating multimedia communications among multiple users [1–5]. The system capacity is a basic problem to research resource management and CAC schemes.

In a CDMA system for multimedia services, each service is specified by QoS requirements such as a target BER and an information data rate. Different types of services are characterized by their different channel quality requirements or different information data rate requirements [5, 6].

In general, different types of services require different received signal power levels, and the amount of interference generated by one service user is different from that generated by another service user. The upper limit for the number of users of a certain service group should be limited by the numbers of users in the other service groups. To fully utilize multimedia CDMA system resources, the system capacity must be identified, and correct tradeoffs are required between the number of users in each service group. Recently, the relationship between the numbers of users in various service groups for a multimedia CDMA has been implicitly addressed [5] and further visualized for a single cell environment [7]. In this chapter, the relationship between the numbers of supportable users in various service groups is investigated for a practical multiple cell environment, and the possibility of using the concept of the capacity plane for resource management design is presented.

This chapter is organized as follows: Following this introduction, the system model is described with the assumptions, and the problem to be analyzed is formulated in Section 2.2. Based on the model, the capacities of CDMA systems for multimedia services in a single cell and a multiple cell environment are evaluated in Sections 2.3 and 2.4, respectively. Finally, concluding remarks are made in Section 2.5.

2.2 System Model and Analysis

The reverse link of single cell and multiple cell systems is considered. To model various services, N user groups are assumed. One group is for voice service, and the other groups are for various data services. Users in one group have the same quality requirement and information data rate requirement. Define the power received by the BS as $S_{v,i}$ for the ith voice user in the voice user group and $S_{d_j,h}$ for the hth user in the data user group j ($j = 1, 2, \ldots, N - 1$), and define the information data rates as R_v for the voice user group and R_{d_j} for the data user group j. For the ith voice user, the received E_b/N_0 is represented as follows [5, 8].

$$\left(\frac{E_b}{N_0}\right)_{v,i} = \frac{W}{R_v} \frac{S_{v,i}}{\sum_{k=1,k\neq i}^{N_v} \alpha S_{v,k} + \sum_{j=1}^{N-1} \sum_{h=1}^{N_{d_j}} S_{d_{j,h}} + I + \eta_0 W} \tag{2.1}$$

where W is the spreading bandwidth; N_v and N_{d_j} represent the number of users in the voice user group and the data user group j in a sector, respectively; α is the voice activity factor; I is the other cell interference; and η_0 is the level of the background noise power spectral density. For the simplicity of the analysis, there are some assumptions:

1. Each BS is assumed to use three ideal directional antennas.
2. The path loss attenuation between the user and the BS is proportional to $10^{\xi/10} r^{-4}$, where r is the distance from the user to the BS and ξ is a Gaussian random variable with zero mean and standard deviation $\sigma = 8$ dB. Fast fading is assumed not to affect the power level.
3. Perfect power control mechanism is assumed.

According to the perfect power control, we have $S_{v,k} = S_v$ and $S_{d_j,h} = S_{d_j}$ for all k and h. From the fact that the background noise η_0 can be negligible compared to the user interference, (2.1) is approximately modified to

$$\left(\frac{E_b}{N_0}\right)_{v} \approx \frac{W}{R_v} \frac{S_v}{\alpha(N_v - 1)S_v + \sum_{j=1}^{N-1} N_{d_j} S_{d_j} + I} \tag{2.2}$$

Similarly, the received E_b/N_0 for the data user group j is

$$\left(\frac{E_b}{N_0}\right)_{d_i} \approx \frac{W}{R_{d_i}} \frac{S_{d_i}}{\alpha N_v S_v + \left(N_{d_i} - 1\right) S_{d_i} + \sum_{j=1, i \neq j}^{N-1} N_{d_j} S_{d_j} + I} \tag{2.3}$$

$$\text{for } j = 1, 2, \ldots, N-1$$

for any certain case of $N_v \neq 0$ and $N_{d_j} \neq 0$. From (2.2) and (2.3), the relation between the received signal powers of user groups is achieved for the case $(E_b/N_0)_v \neq 0$ $(N_v \neq 0)$ and $(E_b/N_0)_{d_j} \neq 0$ $(N_{d_j} \neq 0)$.

$$\left\{(SIR)_v^{-1} + \alpha\right\} S_v = \left\{(SIR)_{d_i}^{-1} + 1\right\} S_{d_i} \tag{2.4}$$

where

$$(SIR)_v = \frac{R_v}{W} \cdot \left(\frac{E_b}{N_0}\right)_v \text{ and } (SIR)_{d_i} = \frac{R_w}{W} \cdot \left(\frac{E_b}{N_0}\right)_{d_i} \tag{2.5}$$

To satisfy the quality requirement, which is one of factors characterizing various services for all user groups, the received E_b/N_0s should be greater than the required E_b/N_0s.

$$\left(\frac{E_b}{N_0}\right)_v \geq \left(\frac{E_b}{N_0}\right)_{v_{req}} \text{ and } \left(\frac{E_b}{N_0}\right)_{d_i} \geq \left(\frac{E_b}{N_0}\right)_{d_{i,req}} \tag{2.6}$$

To satisfy the information data rate requirement for all user groups, the following relations should be satisfied:

$$R_v \geq R_{v_{req}}, R_{d_j} \geq R_{d_{j,req}} \tag{2.7}$$

According to (2.6) and (2.7), the received E_b/N_0s represented in (2.2) and (2.3) are limited as follows:

$$\left(\frac{E_b}{N_0}\right)_{v_{req}} \leq \left(\frac{E_b}{N_0}\right)_v \leq \frac{W}{R_{v_{req}}} \frac{S_v}{\alpha(N_v - 1) S_v + \sum_{j=1}^{N-1} N_{d_j} S_{d_j} + I} \tag{2.8}$$

$$\left(\frac{E_b}{N_0}\right)_{d_{j\,req}} \leq \left(\frac{E_b}{N_0}\right)_{d_j} \leq \frac{W}{R_{d_{j\,req}}} \frac{S_{d_j}}{\alpha N_v S_v + \left(N_{d_j} - 1\right) S_{d_j} + \sum_{\substack{i=1 \\ i \neq j}}^{N-1} N_{d_i} S_{d_i} + I} \tag{2.9}$$

From these equations, the numbers of users, $(N_v, N_{d_1}, N_{d_2}, \ldots, N_{d_{N-1}})$ are upper bounded as follows:

$$\alpha(N_v - 1)S_v + \sum_{j=1}^{N-1} N_{d_j} S_{d_j} + I \le \frac{W}{R_{v_{req}}} \left(\frac{E_b}{N_0}\right)_{v_{req}}^{-1} S_v = (SIR)_{v_{req}}^{-1} S_v \tag{2.10}$$

Applying the relation between the received signal powers of the user groups, as in (2.4)–(2.10), we can derive the relation between the user numbers and the required SIRs.

$$\alpha \frac{N_v}{(SIR)_{v_{req}}^{-1} + \alpha} + \sum_{j=1}^{N-1} \frac{N_{d_j}}{(SIR)_{d_{j\,req}}^{-1} + 1} \le 1 - z \tag{2.11}$$

where

$$z = \frac{I}{S_v} \frac{1}{(SIR)_{v_{req}}^{-1} + \alpha} = \frac{I}{S_{d_j}} \frac{1}{(SIR)_{d_{j\,req}}^{-1} + 1} \tag{2.12}$$

2.3 Single Cell CDMA Capacity

For a single cell system, the other cell interference has no effect on the capacity, and the term z of (2.11) is set to zero. Therefore, (2.11) is simplified to the following equation for a single cell case:

$$\gamma_v N_v + \sum_{i=1}^{N-1} \gamma_{d_i} N_{d_i} \le 1 \tag{2.13}$$

where

$$\gamma_v = \frac{\alpha}{(SIR)_{v_{req}}^{-1} + \alpha} \text{ and } \gamma_{d_i} = \frac{1}{(SIR)_{d_{i\,req}}^{-1} + 1} \tag{2.14}$$

This equation specifies a capacity plane in the N dimensional space. All points $(N_v, N_{d_1}, N_{d_2}, \ldots, N_{d_{N-1}})$ under the hyperplane represent possible numbers of supportable users in voice and data user groups in a sector. In (2.13), total resource amount of the system, the resource amount used by one voice user, and the resource amount used by one data user in the group i correspond to 1, γ_v, and γ_{d_i}, respectively. Equation (2.13) also means that the resources used by users should not exceed total system resource.

Let's consider a system with two user groups, voice and data. The system parameters are shown in Table 2.1. The capacity regions are plotted for several cases. In Figure 2.1, upper limits for the number of users are plotted using several

Table 2.1 Parameters of a CDMA System Supporting Voice and Data Services

Item	*Symbol*	*Value*
Bandwidth	W	1.25 MHz
Voice activity factor	α	0.375
Information data rate for the voice group	R_v	9.6 Kbps
Information data rate for the data group	R_{d_i}	2.4, 4.8, 7.2, and 9.6 Kbps
Quality requirement for the voice group	$\left(\frac{E_b}{N_0}\right)_{v_{req}}$	5 (7 dB)
Quality requirement for the data group	$\left(\frac{E_b}{N_0}\right)_{d_{i,req}}$	12, 10, 5

quality requirements for data user group ($(E_b/N_0)_{d_{req}}$ = 12, 10, and 5). In Figure 2.2, upper limits for the number of users are plotted using several data rates for the data user group (R_d = 9.6, 7.2, 4.8, and 2.4 Kbps). In Figures 2.1 and 2.2, different lines represent the different service cases, and all points (N_v, N_d) under the line represent the possible numbers of supportable users of the voice and data user groups per sector where N_v and N_d are integer. It is observed that the ratio of the system resource used by one voice user to the system resource used by one data user corresponds to the slope of the line, γ_v / γ_d.

Figures 2.1 and 2.2 also show that the user group that requires higher quality or information data rate has a lower limit of the maximum number of users, and this means that the user in that group uses more system resources. As another example, let's consider a system with three user groups. One group is for voice users who have $(E_b/N_0)_{v_{req}}$ = 5 (7 dB) and R_v = 9.6 Kbps. Another group is for data users who have $(E_b/N_0)_{d_1\,req}$ = 10 (10 dB) and R_{d_1} = 9.6 Kbps. The other group is also for data users who have $(E_b/N_0)_{d_2\,req}$ = 10 and R_{d_2} = 4.8 Kbps.

Figure 2.3 shows a three-dimensional capacity plane. As in Figures 2.1 and 2.2, all points (N_v, N_{d_1}, N_{d_2}) under the plane represent the possible numbers of supportable users in the voice and two data user groups, where N_v, N_{d_1}, and N_{d_2} are integers.

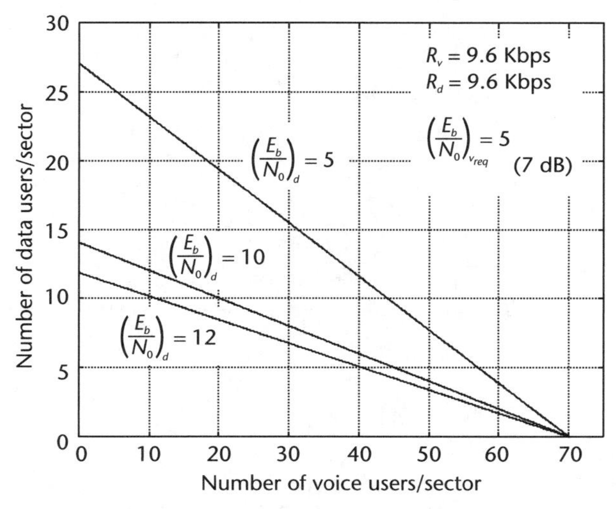

Figure 2.1 Capacity lines for the number of voice users versus the number of data users in a single cell case when $(E_b/N_0)_{d_{req}}$ is given as 12, 10, or 5.

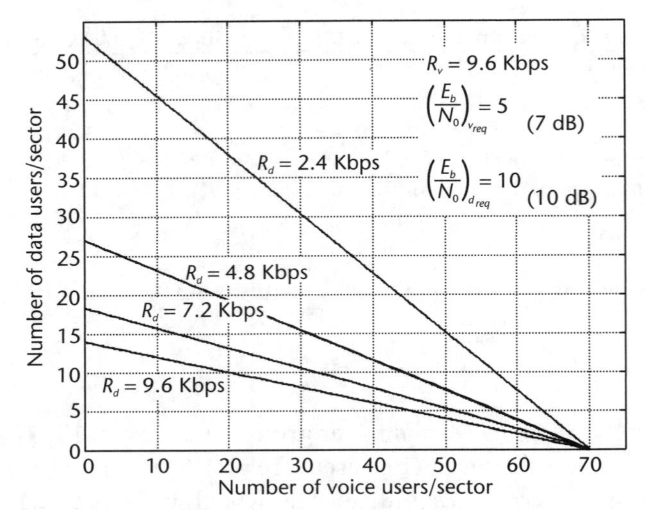

Figure 2.2 Capacity lines for the number of voice users versus the number of data users in a single cell case when R_d is given as 9.6, 7.2, 4.8, or 2.4 Kbps.

The maximum numbers of supportable users are found to be 70 for the voice user group, 14 for data user group 1, and 27 for data user group 2, as in Figure 2.3.

2.4 Multiple Cell CDMA Capacity

For a multiple cell system, users in the other cells generate additional interference compared with a single cell case, where the other users in the same cell generate the interference to the desired user. The effect of the other cell interference on the

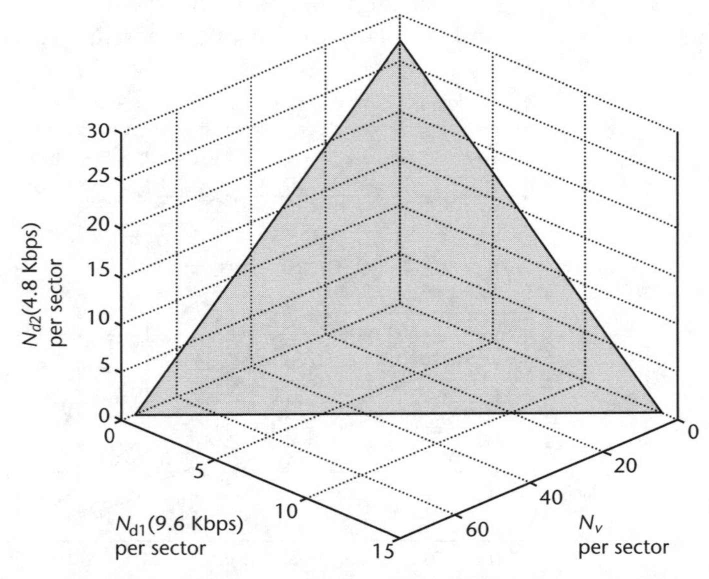

Figure 2.3 Capacity plane for three user groups in a single cell case where $(E_b/N_0)_{v_{req}}$ and R_v are given as 5 and 9.6 Kbps for voice user group, $(E_b/N_0)_{d_1\,req}$ and R_{d_1} are given as 10 and 9.6 Kbps for data user group 1, and $(E_b/N_0)_{d_2\,req}$ and R_{d_2} are given as 10 and 4.8 Kbps for data user group 2.

capacity is included as the term z in (2.11). In the multicell case, it is necessary to characterize the other cell interference I before characterizing z.

The other cell interference in the CDMA system for the voice service has been modeled as a Gaussian random variable [9], where the mean and variance can contribute to characterize the capacity of the system. To analyze the mean and variance, there have been additional assumptions of a uniform distribution of users in the service area, the use of the smallest distance rather than the smallest attenuation to determine home cell and spatial whiteness. Similarly, the other cell interference to the multimedia service environment is also modeled as a Gaussian random variable:

$$I = \iint \left(\phi S_v \rho_v + \sum_{i=1}^{N-1} S_{d_i} \rho_{d_i} \right) \left(\frac{r_m}{r_0} \right)^4 10^{(\xi_0 - \xi_m)/10} \cdot \Phi\left(\xi_0 - \xi_m, \frac{r_0}{r_m} \right) dA \tag{2.15}$$

where ϕ is the voice activity variable, a binomial random variable whose mean is the voice activity factor α. r_0 is the distance from a user in another cell to the desired BS, and r_m is the distance from that user to its BS (see Figure 2.4). m is the BS index,

$$\Phi(\xi_0 - \xi_m, r_0 / r_m) = \begin{cases} 1, & \text{if } (r_m / r_0)^4 10^{(\xi_0 - \xi_m)/10} \leq 1 \\ 0, & \text{otherwise} \end{cases} \tag{2.16}$$

ρ_v is the voice user density, and ρ_{d_i} is the user density in the data user group i.

Following the similar procedure in [9], and assuming the service area is considered up to the second ring—the integral in (2.15) is over the shaded area in Figure

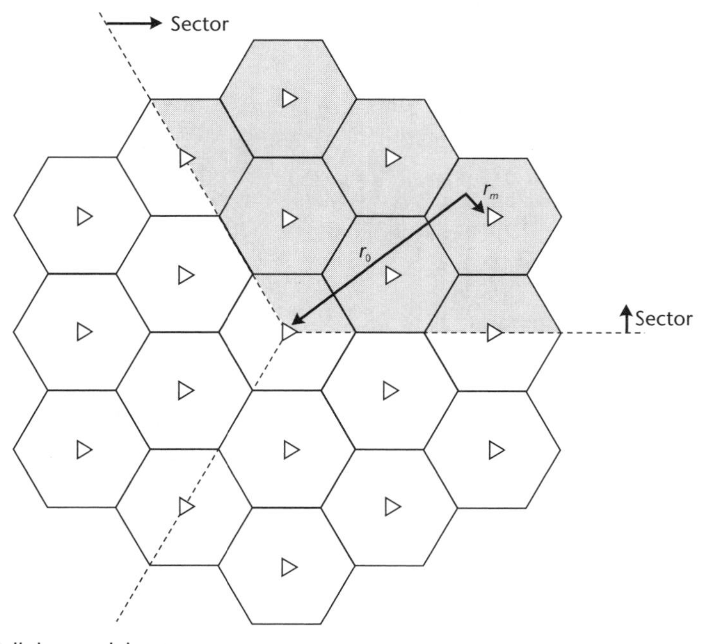

Figure 2.4 Cellular model.

2.4—and there is no overlapping user at the same spatial point, the mean and variance of the other cell interference I are obtained as

$$\begin{aligned} E(I) &\leq 0.247 N_v S_v + 0.659 \sum_{i=1}^{N-1} N_{d_i} S_{d_i} \\ \operatorname{var}(I) &\leq 0.078 N_v S_v^2 + 0.183 \sum_{i=1}^{N-1} N_{d_i} S_{d_i}^2 \end{aligned} \tag{2.17}$$

Using (2.12) and (2.17) to characterize z, z is also modeled as a Gaussian random variable with mean and variance such as

$$\begin{aligned} E(z) &\leq 0.659 \gamma_v N_v + 0.659 \sum_{i=1}^{N-1} \gamma_{d_i} N_{d_i} \\ \operatorname{var}(z) &\leq 0.555 \gamma_v^2 N_v + 0.183 \sum_{i=1}^{N-1} \gamma_{d_i}^2 N_{d_i} \end{aligned} \tag{2.18}$$

For the capacity of a multiple cell CDMA system, (2.11) is used to include the effect of the other cell interference.

$$\gamma_v N_v + \sum_{i=1}^{N-1} \gamma_{d_i} N_{d_i} \leq 1 - z \tag{2.19}$$

By comparing (2.13) with (2.19), we know that total system resource is decreased as much as z due to other cell interference. Assuming that the performance requirements are achieved, P is lower bounded by the required system reliability, which is usually given by 99% [9] such that P is given as like

$$P = \Pr\left\{ \gamma_v N_v + \sum_{i=1}^{N-1} \gamma_{d_i} N_{d_i} \leq 1 - z \right\} \geq 0.99 \tag{2.20}$$

As the random variable z is a Gaussian random variable with mean and variance given in (2.18), (2.20) is easily calculated to be

$$\gamma_v N_v + \sum_{i=1}^{N-1} \gamma_{d_i} N_{d_i} + E(z) + 2.33 \sqrt{\operatorname{var}(z)} \leq 1 \tag{2.21}$$

where $E(z)$ and $\operatorname{var}(z)$ are the functions of N_v and N_{d_i}. Thus, compared with the results of the single cell system, the resource used by a voice user is greater than γ_v (for voice user in a single cell system) and the resource used by a data user in group i is also greater than γ_{d_i} (for group i data user in a single cell system), while total system resource (regarded as 1) is same as that of the single cell system.

For example, let's consider a system with two user groups, including one voice user group and one data user group. The system parameters in Table 2.1 are also used.

Figure 2.5 shows the upper bounds for the number of voice users versus the number of data users for several $(E_b/N_0)_{req}$ values of a data service group.

Figure 2.6 also shows the upper bounds for the number of voice users versus the number of data users for several bit rate constraints for a data user group. As another example, let's consider a system with three user groups as with the previous single cell case.

Figure 2.7 shows the three-dimensional capacity region for the multicell case, where the maximum possible numbers of users are found to be 36 for the voice user group, 5 for data user group 1, and 12 for data user group 2.

Particularly, a vertex value of (N_v, N_{d_1}, N_{d_2}), (36.08, 0, 0) corresponds to the voice-only user capacity of the IS-95 CDMA system.

2.5 Conclusions

In this chapter, the capacities of single cell and multiple cell CDMA systems supporting multimedia services have been evaluated. Both capacities are confined by a deterministic hyperplane (namely, a capacity plane), whose dimension is determined by the number of service groups. The amount of system resources required by one service user is compared with that required by another service user based on the slope of capacity lines in figures that are presented in Sections 2.3 and 2.4. As expected, the user who requires higher quality or a higher information data rate uses more system resources. Comparing the capacity of a single cell case with that of a multiple cell case, we know that the capacity of the multiple cell case is confined by a lower hyperplane than that of the single cell system due to the effect of the other cell interference.

The concept of the capacity plane can be used for CAC schemes in multimedia service environments. For example, when a new user requests a service, the system resource required by the user can be expected. If the system resource required by the user is smaller than the remaining system resource, then the user is accepted.

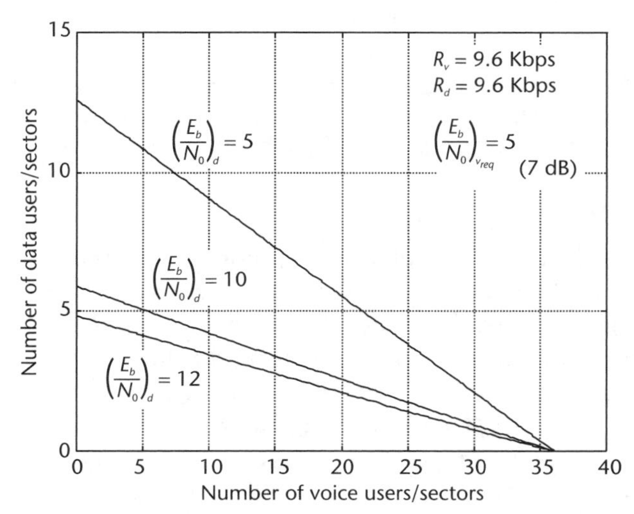

Figure 2.5 Capacity lines for the number of voice users versus the number of data users in a multiple cell case.

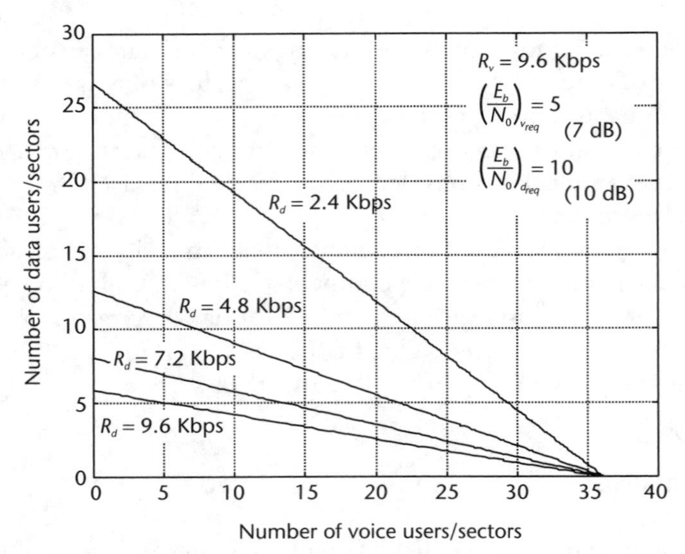

Figure 2.6 Capacity lines for the number of voice users versus the number of data users in a single cell case.

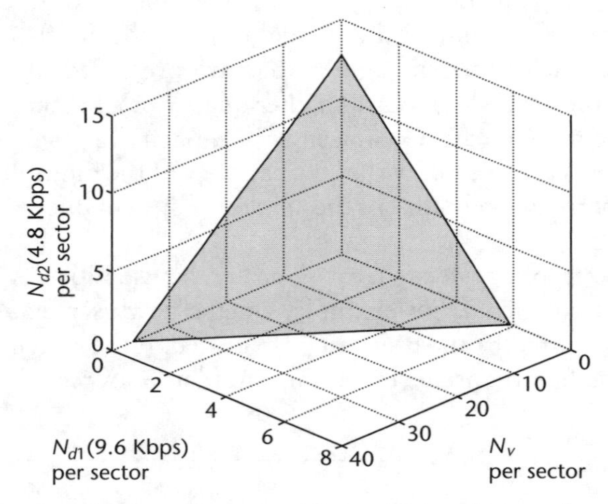

Figure 2.7 Capacity plane for three user groups in a multiple cell case, where $(E_b/N_0)_{v_{req}}$ and R_v are given as 5 and 9.6 Kbps for voice user group, $(E_b/N_0)_{d_{1req}}$ and R_{d_1} are given as 10 and 9.6 Kbps for data user group 1, and $(E_b/N_0)_{d_{2req}}$ and R_{d_2} are given as 10 and 4.8 Kbps for data user group 2.

However, if the required system resource is greater than the remaining system resource, then the user is blocked [5]. For such applications, in this book, we will utilize the evaluated capacity plane as a reference for the threshold for CAC when evaluating the corresponding Erlang capacity of CDMA systems. Particularly in Chapters 7 through 10, we tackle such applications to evaluate the Erlang capacity.

In addition, the capacity plane can be used for system resource management [10]. For example, if current users in the system do not use all of the system resources, the remaining system resources may be allowed to go to the current users to increase the throughput or the quality until a new user requests a service and resource allocation is newly made to accept the user. On the other hand, some kinds of smart blocking/acceptance mechanism [11] can be devised where we can accept a user with diminished but tolerable QoS, even though the remaining system resources are not enough to accept the request call. For such applications of resource allocation, in Chapter 5 we will present an efficient resource allocation scheme to fully utilize the remaining resources in the system with which we can find the optimum set of data rates for concurrent users and further maximize the system throughput while satisfying the minimum QoS requirements of each user.

References

[1] Wu, J., and R. Kohno, "Wireless Multi-Media CDMA System Based on Transmission Power Control," *Proc. of IEEE International Symposium on Personal, Indoor and Mobile Radio Communications*, 1995, pp. 36–40.

[2] Gejji, R. R., "Mobile Multimedia Scenario Using ATM and Microcellular Technologies," *IEEE Trans. on Vehicular Technology*, 1994, pp. 699–703.

[3] McTiffin, M. J., et al., Mobile Access to an ATM Network Using a CDMA Air Interface," *IEEE Journal on Selected Areas in Communications*, 1994, pp. 900–908.

[4] Yang, W. B., and E. Geraniotis, "Admission Policies for Integrated Voice and Data Traffic in CDMA Packet Radio Networks," *IEEE Journal on Selected Areas in Communications*, 1994, pp. 654–664.

[5] Sampath, A., P. S. Kumar, and J. M. Holtzman, "Power Control and Resource Management for a Multimedia CDMA Wireless System," *IEEE Proc. of International Symposium on Personal, Indoor, and Mobile Radio Communications,* 1995, pp. 21–25.

[6] Wu, J. S., and J. R. Lin, "Performance Analysis of Voice/Data Integrated CDMA System with QoS Constraints," *IEICE Trans. on Communications*, Vol. E79-B, 1996, pp. 384–391.

[7] Paulrajan, V. K., J. A. Roberts, and D. L. Machamer, "Capacity of a CDMA Cellular System with Variable User Data Rates," *Proc. of IEEE Global Telecommunications Conference,* 1996, pp. 1458–1462.

[8] Viterbi, A. M., and A. J. Viterbi, "Erlang Capacity of a Power-Controlled CDMA System," *IEEE Journal on Selected Areas in Communications,* 1993, pp. 892–900.

[9] Gilhousen, K. S., et al., "On the Capacity of a Cellular CDMA System," *IEEE Trans. on Vehicular Technology,* 1991, pp. 303–312.

[10] Yang, J., et al. "A Dynamic Resource Allocation Scheme to Maximize Throughput in a Multimedia CDMA System," *IEEE Proc. of Vehicular Technology Conference*, 1999, pp. 348–351.

[11] Ko, G., A. Ahmad, and K. Kim, "Analysis of a Variable Rate Access Control Algorithm in Integrated Voice/Data DS-CDMA Networks," *Proc. of IWTS*, 1997, pp. 133–138.

CHAPTER 3

Sensitivity Analysis in CDMA Systems

In CDMA systems, the number of simultaneous users occupying resources should be limited so that an appropriate level of communication quality can be maintained. In this aspect, CAC plays a very important role in CDMA systems because it directly controls the number of users. CAC schemes are usually based on a threshold mechanism whose purpose is to ensure that the performance of users in the system satisfies their specified QoSs. In particular, CACs for the CDMA system can be classified into two schemes: interference-based CAC (ICAC) and number-based CAC (NCAC) [1]. The NCAC admits a new connection if total number of existing connections in the system is less than a predefined value, while the ICAC admits a new connection if total interference in the system is less than a certain threshold. As a reference to such thresholds for CAC in CDMA systems, one of the capacity bounds explained in previous sections can be utilized.

In practice, however, even if a fixed frequency band is used in a cell, the capacity bounds may vary with the loading of home and neighboring cells, mainly because co-channel interference changes according to the loading. For the design of robust and stable CAC schemes, it is important to consider the effect of the disturbance of system parameters on the threshold for CAC schemes, which directly corresponds to the effect of disturbance of the capacity parameters on the system capacity.

A typical way to quantitatively describe the change in the system capacity due to the variation of system capacity parameters is the sensitivity analysis, which relates the elements of the set of the parameter deviations to the elements of the set of the parameter-induced errors of the system function. Such sensitivity analysis has been applied to many system analyses for:

1. Guiding future research by highlighting the most important system parameter;
2. Estimating parameters by obtaining the combination of system parameters that leads to optimum system operation point with respect to the system operator;
3. Evaluating the magnitude of the effect of system parameters errors on the system performance.

In most cases, imperfections encountered in the CDMA systems are due to imperfect power control. The imperfection effect due to imperfect power control on the reverse link capacity of a CDMA system was studied in many papers [2–4]. In addition, the system reliability, defined as the predetermined value of probability that the received signal-to-interference ratio (SIR) is larger than the required SIR, is one of the most important system parameters, as the reverse link capacity is usually limited by a prescribed lower bound of system reliability. In [2], traffic capacity estimation under the power control imperfections in conjunction with the system reliability was presented through simulation. Furthermore, a theoretical analysis of the effect of system reliability on the reverse link capacity was implicitly presented [3]. However, none of these works [2–4] present analytical close-form expression of system reliability on system capacity. Furthermore, only voice-oriented CDMA systems are discussed in the previous works.

As stated in the previous section, CDMA system capacity can be expressed as a function of such system parameters as the required E_b/N_0, traffic activity factor, processing gain, and frequency reuse factor. In addition, the sensitivity of respective parameters on CDMA system capacity can afford a proper measure to design CAC schemes. However, as an example of sensitivity analysis in CDMA systems, in this chapter we focus on the sensitivity of system capacity with respect to system reliability, such that the effects of system reliability as well as the imperfection due to the imperfect power control on the CDMA capacity are considered explicitly through sensitivity analysis. Further, an accurate, simple analytical close-form expression for the limitation of the capacity is shown. However, it is noteworthy that even though only the sensitivity of system capacity with respect to system reliability is presented, the sensitivity of other parameters on the CDMA system capacity can be easily evaluated with the presented analysis method.

3.1 System Model and System Capacity

Let's consider the reverse link of multicell CDMA systems where K user groups are assumed to model various services in the multimedia environment. One group is for voice service, and the others are for various data services. Users in the same group have the same information data rate requirement R, $R \geq R_{req}$ and system reliability requirement $\beta\%$, $P_r(SIR \geq SIR_{req}) = \beta\%$ where $SIR_{req} = (E_b/N_0)_{req} \cdot R_{req}/W$. The received SIR of each user depends on the power control mechanism that attempts to equalize the performance of all users. It is well known to be approximately log-normally distributed with a standard deviation 0.5–2 dB. Furthermore, it is assumed that the allocated frequency bandwidth W, the standard deviation of the received SIR σ_x, and the system reliability $\beta\%$ are the same for all service groups.

To satisfy the requirements of all users, the numbers of users in the system are confined by following equation, which was derived in [5].

$$\gamma_v N_v + \sum_{i=1}^{K-1} \gamma_{d_i} N_{d_i} \leq 1 \tag{3.1}$$

where

$$\gamma_v = \frac{\alpha}{\frac{W}{R_{v_{req}}}\left(\frac{E_b}{N_o}\right)_{v_{req}}^{-1}\left\langle\frac{1}{1+f}\right\rangle 10^{\frac{Q^{-1}(\beta)}{10}\sigma_x - 0.012\sigma_x^2} + \alpha}$$

$$\gamma_{d_i} = \frac{1}{\frac{W}{R_{d_{i\,req}}}\left(\frac{E_b}{N_o}\right)_{d_{i\,req}}^{-1}\left\langle\frac{1}{1+f}\right\rangle 10^{\frac{Q^{-1}(\beta)}{10}\sigma_x - 0.012\sigma_x^2} + 1}$$

N_v and N_{d_i} denote the number of users in the voice user group and the ith data user group, respectively, and r_v and r_{d_i} can be defined as the normalized effective bandwidth of voice and data user in the ith data group, respectively. Q^{-1} is the inverse Q-function where Q-function is defined as $Q(x) = \int_{-\infty}^{x}\left(1/\sqrt{2\pi}e^{-y^2/2}\right)dy$. Equation (3.1) means that the numbers of users in the system, (N_v, N_{d_1}, N_{d_2}, ..., $N_{d_{K-1}}$), are limited in the range that the sum of the normalized effective bandwidth of active users of each service group does not exceed the unit.

From (3.1), we can look at several variables that determine CDMA capacity.

- W is the spreading bandwidth.
- $R_{i_{req}}$ for $i = v, d_1, \ldots, d_{K-1}$ is the required information data rate.
- $(E_b/N_0)_{i_{req}}$ for $i = v, d_1, \ldots, d_{K-1}$ is the required bit energy-to-interference power spectral density ratio.
- f is the other cell interference factor with which the interference contribution from other cells relative to the carrier on the serving cell can be considered.
- α is the voice activity factor, so $\alpha = 1$ represents channels that are always on, and $\alpha = 2/3$ represents channels that are powered off one-third of the time.
- β% is system reliability, which is defined as the predetermined value of probability that the received SIR is larger than the required SIR.
- σ_x is the effect of power-control delays and errors, so perfect power control is $\sigma_x = 0$ dB, and values less than 1 reflect lower performance. In particular, the quantity of $10^{\frac{Q^{-1}(\beta)}{10}\sigma_x - 0.012\sigma_x^2}$ indicates the effect of the imperfect power control error and the system reliability β% on the system capacity. It is noteworthy that γ_v and γ_{d_i} have a similar form of (2.14) when the imperfect power control error σ_x goes to zero, which is mainly because (2.14) has been derived under the condition of the perfect power control.

In practice, the capacity bounds may change with variations of the system capacity parameters. Particularly when the capacity bound is utilized for the threshold of the CAC scheme, it is important to consider the effect of the disturbance of system parameters on system capacity. Imperfections encountered in CDMA systems in most cases are due to imperfect power control error, the effect of which on CDMA capacity is practically linked to system reliability. In this chapter we focus on quantitatively describing the change of system capacity due to the disturbance of power control error, with consideration of system reliability through sensitivity analysis, as an example of sensitivity analysis in CDMA systems.

3.2 The Significance and Definitions of Sensitivity Analysis

A typical way to quantitatively describe the change in the system capacity due to the variation of system capacity parameters is the sensitivity analysis, which relates the elements of the set of the parameter deviations to the elements of the set of the parameter-induced errors of the system function. Such sensitivity analysis has been applied by many system analyses for:

1. Guiding future research by highlighting the most important system parameter;
2. Estimating parameters by obtaining the combination of system parameters that leads to optimum system operation point with respect to the system operator;
3. Evaluating the magnitude of the effect of system parameters error on system performance.

Before applying the sensitivity analysis to our case, we present the basic definitions and significance of sensitivity analysis in the next section.

3.2.1 The Significance of Sensitivity Analysis

The sensitivity of a system to variations of its parameters is one of the basic aspects in the treatment of systems. The question of parameter sensitivity particularly arises in the fields of engineering where mathematical models are used for the purposes of analysis and synthesis. In order to be able to give a unique formulation of the mathematical problem, the mathematical model is usually assumed to be known exactly. This assumption is unrealistic because there is always a certain discrepancy between the actual system and its mathematical model. For this reason, the results of mathematical syntheses need not necessarily be practicable. They may even be very poor if there are considerable parameter deviations between the real system and the mathematical model. This is of particular importance if optimization procedures are involved because it is in the nature of optimization to make extreme a certain performance index for the special set of parameters. Furthermore, there are many other problems where sensitivity considerations are either useful or mandatory. Some examples are the applications of gradient methods, adaptive and self-learning systems, the design of insensitive and suboptimal control systems, the determination of allowed tolerance in the design of networks, the calculation of optimal input signals for parameter identification, and analogy and digital simulation of dynamic systems.

3.2.2 Basic Definitions of Sensitivity

There are several ways to define quantities for the characterization of the parameter sensitivity of a system. Here, some definitions are summarized. Let the behavior of the system be characterized by a quantity $\mathrm{C} = \zeta(\alpha)$, called a system function, which is a function of the parameter vector $\alpha = [\alpha_1\,\alpha_2 \ldots \alpha_r]^T$. Let the nominal parameter vector be denoted by $\alpha_0 = [\alpha_{10}\,\alpha_{20} \ldots \alpha_{r0}]^T$ and the nominal system function by $\mathrm{C}_0 = \zeta(\alpha_0)$. Then, under certain continuity conditions, the following general definitions hold [6].

Definition 3.2.2-1: Absolute Sensitivity Function

$$S_{\alpha_j}^{\mathrm{C}} \equiv \frac{\partial \mathrm{C}}{\partial \alpha_j}\Big|_{\alpha_0} = S_{\alpha_j}^{\mathrm{C}}(\alpha_0) \text{ for } j = 1, 2, \ldots r \tag{3.2}$$

The subscript α_0 indicates that the partial derivative expressed by ∂ is taken at nominal parameter values.

Definition 3.2.2-2: Parameter-Induced Error of the System Function

$$\Delta\zeta \equiv \sum_{i=1}^{r} S_{\alpha_j}^{\mathrm{C}} \Delta\alpha_j \tag{3.3}$$

Definition 3.2.2-3: Maximum Error of the System Function

$$|\Delta\zeta| \equiv \sum_{i=1}^{r} \left|S_{\alpha_j}^{\mathrm{C}}\right| \left|\Delta\alpha_j\right| \tag{3.4}$$

The vertical bars in combination with a vector shall indicate that the absolute values of the elements of the corresponding vector are to be taken.

Definition 3.2.2-4: Relative (Logarithmic) Sensitivity Function

$$\overline{\mathbf{S}}_{\alpha_j}^{\mathrm{C}} \equiv \frac{\partial \ln \mathrm{C}}{\partial \ln \alpha_j}\Big|_{\alpha_0} = \overline{\mathbf{S}}_{\alpha_j}^{\mathrm{C}}(\alpha_0) \text{ for } j = 1, 2, \ldots r \tag{3.5}$$

Note that lnC means the vector of the logarithms of the elements of **C**. Hence, $\partial \ln \mathrm{C} = \left[\partial C_1 / C_1 \ \partial C_2 / C_2 \cdots \partial C_n / C_n\right]^T$. The ith element of $\overline{\mathbf{S}}_{\alpha_j}^{\mathrm{C}}$ can be expressed by

$$\overline{S}_{\alpha_j}^{C_i} = \frac{\partial C_i / C_i}{\partial \alpha_j / \alpha_j} = S_{\alpha_j}^{C_i} \frac{\alpha_{j0}}{\zeta_{i0}} \tag{3.6}$$

where $S_{\alpha_j}^{C_i}$ is the ith element of the absolute sensitivity function $\overline{\mathbf{S}}_{\alpha_j}^{\mathrm{C}}$.

Definition 3.2.2-5: Relative Error of the System Function

The ith element of the relative error of the system function is defined as

$$\frac{\Delta C_i}{\zeta_{i0}} \equiv \sum_{j=1}^{r} \overline{S}_{\alpha_j}^{C_i} \frac{\Delta \alpha_j}{\alpha_{j0}}, \quad i = 1, 2, \ldots, n \tag{3.7}$$

Definition 3.2.2-6: Maximum Relative Error of the System Function

The ith element is defined as

$$\left|\frac{\Delta C_i}{\xi_{i0}}\right| \equiv \sum_{j=1}^{r} \left|\overline{S}_{\alpha_j}^{C_i}\right| \left|\frac{\Delta \alpha_j}{\alpha_{j0}}\right|, \quad i = 1, 2, \ldots, n \tag{3.8}$$

3.3 Sensitivity of System Capacity with Respect to System Reliability in CDMA Cellular Systems

For sensitivity analysis in CDMA systems, a capacity equation that has been driven in many other papers [5, 7–10] can be used. Here, we adopt the result of [5], which includes the effects of imperfect power control error and system reliability on the system capacity whose main result is described in (3.1).

In order to consider the effect of system reliability on the numbers of users in the all service groups simultaneously, equivalent telephone (or voice) capacity (ETC) is specified as a capacity unit, which is defined as the equivalent number of telephone (or voice) channels available in the reverse link [11]. Noting that ETC is the capacity equivalent to the number of voice users, in our case, we have ETC as in (3.9) by referring (3.1) and considering the normalized effective bandwidth of each service group.

$$\hat{C}_{ETC} \equiv N_v + \sum_{i=1}^{K-1} \frac{\gamma_{d_i}}{\gamma_v} N_{d_i} \tag{3.9}$$

For sensitivity analysis, the relative sensitivity in (3.6) is here adopted among various definitions because it provides a unitless measure over a wide range of parameters, and further we set $\hat{C}_{ETC}$ as a system function, C and β as parameter vectors, with α to follow the notations of (3.6). Then, the sensitivity of ETC with respect to the system reliability $\beta\%$ is written as

$$\begin{aligned} \overline{S}_{\beta}^{\hat{C}_{ETC}} &\equiv \frac{\partial \ln \hat{C}_{ETC}}{\partial \ln \beta} \Big|_{\beta_o} \\ &= -\frac{\sigma_{x_o}}{10 \ln 10} \beta_o \sqrt{2\pi} \exp\left(Q^{-1}(\beta_o)^2\right) \end{aligned} \tag{3.10}$$

where the subscript of "o" denotes the normal value of each system parameter for the system operation.

Note that the sensitivity of ETC with respect to the system reliability $\beta\%$ is expressed in terms of system reliability and the standard deviation of the received SIR. It means that the variation of the received SIR degrades the performance of the system capacity and the degree of degradation depends on the system reliability.

Figure 3.1 depicts the sensitivity of ETC with respect to system reliability as a function of the standard deviation of the received SIR and the system reliability. Here, the IS-95-type CDMA system supporting voice and data services is considered for a numerical example. The interference caused by other users is modeled as an

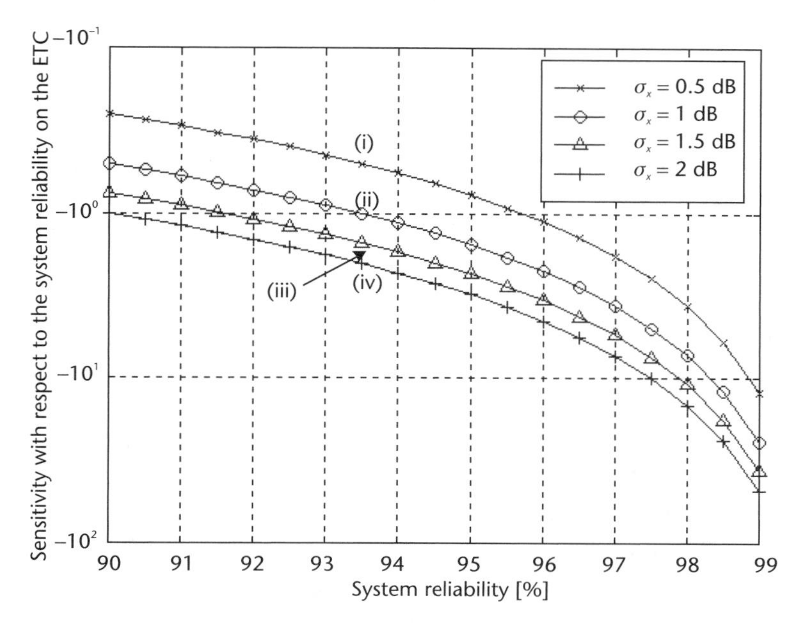

Figure 3.1 Sensitivity with respect to system reliability on system capacity: (a) the standard deviation of the received SIR = 0.5 dB, (b) the standard deviation of the received SIR = 1 dB, (c) the standard deviation of the received SIR = 1.5 dB, and (d) the standard deviation of the received SIR = 2 dB.

additive white Gaussian noise. The amount of interference caused in other cells is assumed to be 0.45 times the interference caused in the home cell. The spreading bandwidth is 1.2288 MHz. The voice activity factor is 3/8. The adequate BER performances of voice and data traffic are required $BER_v \leq 10^{-3}((E_b/N_0)_{v_{req}} = 7$ dB) and $BER_d \leq 10^{-5}((E_b/N_0)_{d_{req}} = 10$ dB), respectively. Figure 3.1 shows that system capacity is very sensitive to system reliability. More specifically, sensitivity with respect to system reliability on system capacity, especially between 95% and 99%—the range in which we are interested—has a value ranging from 5 to 50 when $\sigma_x = 2$ dB, which is relatively high compared with the sensitivity of parameters such as the required E_b/N_0, traffic activity factor, the processing gain, and frequency reuse factor, all of which is near 1 [12]. Figure 3.1 also indicates that a greater variation of the received SIR results in greater sensitivity of system reliability on the ETC. Hence, the limitation of the ETC caused by the system reliability is more increased at the high variation of the received SIR.

Another important task is to estimate the magnitude of change in the system capacity due to the disturbance of the system parameters. In sensitivity theory, it is easy to calculate the change in the system behavior due to the given parameter variations when the sensitivity is known. If the system reliability is given as β_o% and the disturbance of the system reliability is $\Delta\beta$%, then in our case the change of the system capacity caused by the disturbance of the system reliability is given as

$$\Delta\hat{C}_{ETC} = \overline{S}_{\beta}^{\hat{C}_{ETC}} \frac{\Delta\beta}{\beta_o} \hat{C}_{ETC_o} \tag{3.11}$$

For example, 1% change of the system reliability (i.e., from 98% to 99%) results in the capacity reduction of 2.2 when the system reliability and the standard deviation are given as $\beta_o\% = 98\%$ and $\sigma_{x_o} = 1$ dB. In Figure 3.2, curve (c) shows the capacity line degraded by a 1% change of system reliability while the normal value of the system reliability varies between 90% and 99%. Note that the same variation of the system reliability induces a higher change of the system capacity at the high system reliability than at the low system reliability.

Furthermore, the effect of system reliability on system capacity caused by the disturbance of the standard variation of the received SIR is also considered because the limitation of system capacity by system reliability is related with the standard deviation of the received SIR. It is given as $\Delta\hat{C}_{ETC} = \overline{S}_{\sigma_x}^{\hat{C}_{ETC}} \frac{\Delta\sigma_x}{\sigma_{x_o}} \hat{C}_{ETC_o}$ where $\overline{S}_{\sigma_x}^{\hat{C}_{ETC}} = \frac{Q^{-1}(\beta_o)}{10\ln 10}\sigma_{x_o} - (0.024/\ln 10)\sigma_{x_o}^2$. Similarly, the 10% change of the standard deviation of the received SIR, $\Delta\sigma_x = 0.1$ dB, results in $\hat{C}_{ETC} = 0.3$. In Figure 3.2, curve (b) shows the change of ETC caused by the 10% change of the standard deviation of the received SIR, as the normal value of system reliability varies between 90% and 99%. For certain normal values of system reliability, the 1% variation of the system reliability from the normal value induces much higher change in the system capacity than the 10% variation of the standard deviation of the received SIR. This is because system capacity is more sensitive to system reliability than the variation of the received SIR.

The capacity from the viewpoint of the number of voice users has been considered so far. However, the definition of the capacity to consider the number of users in the ith data group can be changed. Based on (3.1), it is also clear that one data user in the ith data group is equivalent to K_i voice users where $K_i \equiv \frac{\gamma_{d_i}}{\gamma_v}$. Then, the new

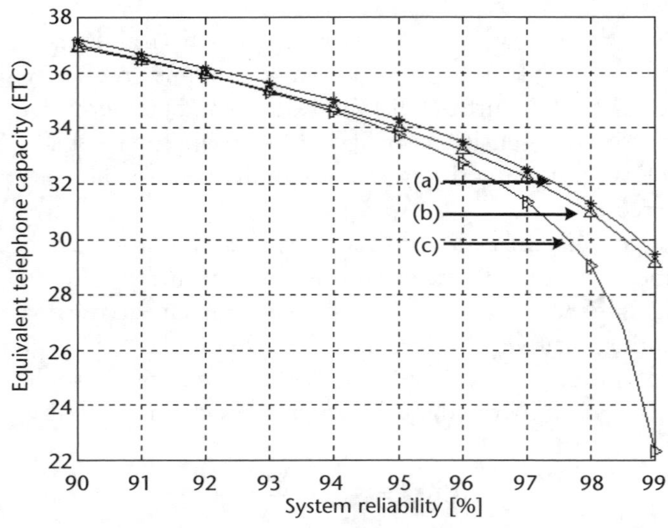

Figure 3.2 Change of ETC induced by the parameter error: (a) the capacity line at the normal values, (b) the capacity line degraded by a 10% variation of the received SIR from the normal value, and (c) the capacity line degraded by a 1% variation of the system reliability from the normal value.

capacity with respect to the number of the ith data users, C_{ED_iC}, can be expressed as $\hat{C}_{ETC}/K_i$ where the subscript ED_iC means the equivalent data capacity with respect to the ith data group. Using the sensitivity quotient rule, sensitivity with respect to system reliability in the number of ith data users $\overline{S}_{\beta}^{C_{ED_iC}}$ can be given as $\overline{S}_{\beta}^{\hat{C}_{ETC}} - \overline{S}_{\beta}^{K_i}$ for all i where $\overline{S}_{\beta}^{K_i} = \frac{\partial \ln K_i}{\partial \ln \beta}|_{\beta_0}$ [6].

For practical values of system parameters, K_i does not change due to system reliability, such that $\overline{S}_{\beta}^{K_i}$ can be negligible. Furthermore, intuitively, the sensitivity of the system capacity is a relation between the relative change of the system capacity and the system reliability. Hence, $\overline{S}_{\beta}^{\hat{C}_{ETC}}$ and $\overline{S}_{\beta}^{C_{ED_iC}}$ should have similar values. For this reason, $\overline{S}_{\beta}^{\hat{C}_{ETC}}$ has been considered only as a practical measure of the sensitivity of system reliability on system capacity.

3.4 Conclusion

As an example of sensitivity analysis in CDMA systems, in this chapter an accurate and simple analytical close-form expression on the limitation of system capacity due to system reliability is shown for the reverse link of multimedia CDMA systems. As a result, the effect of system reliability on system capacity can be expressed in terms of system reliability and the standard deviation of the received SIR. The effect of system reliability on system capacity is proportional to the variation of the received SIR. In a numerical example, sensitivity with respect to system reliability on system capacity, especially in the range between 95% and 99%, has a value ranging from 5 to 50 when $\sigma_x = 2$ dB, which is relatively high compared with the sensitivity of parameters such as the required E_b/N_0, traffic activity factor, processing gain, and frequency reuse factor, all of which have a value of about 1 [12]. Furthermore, an estimated value of the magnitude of the change in system capacity due to the disturbance of system reliability and the standard deviation of the received SIR was presented.

References

[1] Ishikawa, Y., and N. Umeda, "Capacity Design and Performance of Call Admission Control in Cellular CDMA Systems," *IEEE Journal on Selected Areas in Communications,* 1997, pp. 1627–1635.

[2] Kudoh, E. "On the Capacity of DS/CDMA Cellular Mobile Radios Under Imperfect Transmitter Power Controls," *IEICE Trans. Commun.*, 1993, pp. 886–893.

[3] Prasad, R., M. Jansen, and A. Kegel, "Capacity Analysis of a Cellular Direct Sequence Code Division Multiple Access System with Imperfect Power Control," *IEICE Trans. Commun.*, 1993, pp. 894–905.

[4] Ariyavisitakul, S., and L. Chang, "Signal and Interference Statistics of a CDMA System with Feedback Power Control," *IEEE Trans. on Communications*, 1993, pp. 1626–1634.

[5] Koo, I., et al., "A Generalized Capacity Formula for the Multimedia Traffic," *Proc. of Asia-Pacific Conference on Communications*, 1997, pp. 46–50.

[6] Frank, P., *Introduction to System Sensitivity Theory*, New York: Academic Press, 1978.

[7] Gilhousen, K. S., et al., "On the Capacity of a Cellular CDMA System," *IEEE Trans. on Vehicular Technology*, 1991, pp. 303–312.

[8] Sampath, A., P. S. Kumar, and J. M. Holtzman, "Power Control and Resource Management for a Multimedia CDMA Wireless System," *IEEE Proc. of International Symposium on Personal, Indoor, and Mobile Radio Communications,* 1995, pp. 21–25.

[9] Paulrajan, V. K., J. A. Roberts, and D. L. Machamer, "Capacity of a CDMA Cellular System with Variable User Data Rates," *Proc. of IEEE Global Telecommunications Conference*, 1996, pp. 1458–1462.

[10] Yang, Y. R., et al., "Capacity Plane of CDMA Systems for Multimedia Traffic," *IEE Electronics Letters,* 1997, pp. 1432–1433.

[11] Cheung, J., M. Beach, and J. McGeehan, "Network Planning for Third Generation Mobile Radio Systems," *IEEE Commun. Mag.*, 1994, pp. 54–59.

[12] Koo, I., et al, "Sensitivity Analysis for Capacity Increase on the DS-CDMA System," *Proc. of JCCI*, 1997, pp. 447–451.

CHAPTER 4

Effect of Traffic Activity on System Capacity

Drs. J. Yang and K. Kim

It is well known that CDMA systems are interference limited, which implies that the multiaccess interference (MAI) is a key parameter that governs system performance and capacity. Fading and the random activity of users are two fundamental ingredients of the MAI. In this chapter, we focus primarily on investigating the effect of traffic activity on the capacity of CDMA systems, based on ON/OFF traffic models.

In CDMA systems, the interference can be suppressed by monitoring the traffic activity of users, which corresponds to improving the system capacity because CDMA systems are interference limited. The simplest way to take into account the effect of traffic activity on system capacity is to consider the long-term average interference, which simplifies the random characteristics of traffic activity into the mean of traffic activity. A more practical way is to statistically consider the fluctuation of interference due to the traffic activity by modeling the traffic activity as a binomial random variable. In this chapter, the capacity of a CDMA system supporting multiclass services with ON/OFF activity is analyzed based on the latter way, and the corresponding capacity is compared with the capacity analyzed the former way. The influence of traffic activity on the system capacity is further investigated under the same transmission rate and under the same average rate. According to the traffic activity factor, the average rate changes under the same transmission rate, while the transmission rate changes under the same average rate. From the investigation under the same average rate, it is shown that the system capacities for users with different traffic activities are different from each other, even though the average amount of information data to be transmitted in a certain time duration is same.

4.1 Introduction

As the capacity of a CDMA system is interference limited, any reduction of the interference corresponds to improve the system capacity [1]. One technique to reduce the interference is to operate the system in a DTX mode for traffic with ON/OFF traffic activity [2].

Figure 4.1 shows an example of time-based ON/OFF trajectory of traffic activity. In the DTX, the transmission can be suppressed when there is no data to be sent (i.e., the interference can be suppressed when the user is on an idle, or OFF, state). The simplest way to consider the reduction of interference due to traffic activity in capacity analysis is to consider the long-term average interference, where the

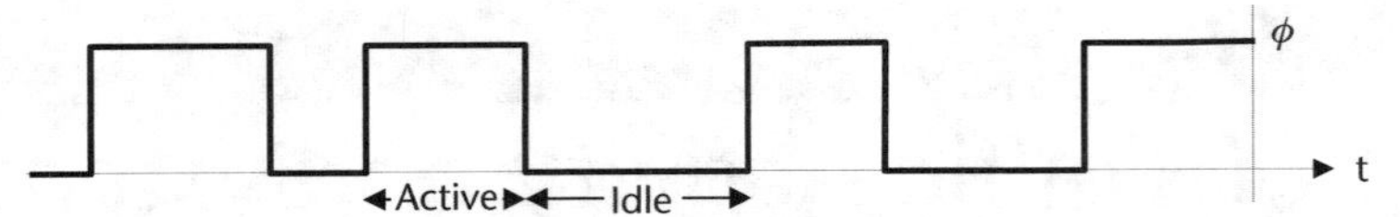

Figure 4.1 An example of time-based ON/OFF trajectory of traffic activity.

random characteristics of traffic activity is assumed to be represented by the mean of traffic activity, called the *traffic activity factor* [1, 3, 4]. For instance, the interference was assumed to be averaged out and reduced by a factor of the reciprocal of the voice traffic activity factor for a preliminary capacity analysis for a voice-only CDMA system [1]. In [3, 4], the same assumption was used to analyze the capacity of a voice/data CDMA system. However, because the probability that the interference exceeds the average interference cannot be negligible, a more practical way is needed to statistically consider the fluctuation of the interference due to the traffic activity. Thus, in this chapter, we model the traffic activity as a binomial random variable [1, 5]. For convenience, we name the former and latter ways as the average interference limited method (AILM) and the statistical interference limited method (SILM), respectively. In [5], the capacity of one service was assumed to have a linear relationship with that of the other service for a voice/data CDMA system, where the capacity of each service was analyzed independently with the SILM. In this chapter, we more precisely analyze the system capacity by considering different services together and further extend this analysis to a CDMA system supporting multiclass services. The capacity analyzed with the SILM is also compared to that analyzed with the AILM.

Because the investigation on the effect of traffic activity was originated in a voice-only system with the same transmission rate, most studies have focused on capacity improvement for several specific values of the traffic activity factor under the same transmission rate [1, 6–8]. However, in the system supporting multiclass services, each service group has different transmission rates, or different activity factors. Another investigation on the effect of traffic activity was performed for several specific values of the activity factor under the same average rate [4]. Under the same average rate, the transmission rate changes according to the traffic activity factor. In this chapter, we investigate the overall dependency of system capacity on traffic activity under the same transmission rate and under the same average rate.

This chapter is organized as follows: Following this introduction, the system capacity is analyzed with the SILM and compared to the capacity analyzed with the AILM with respect to the outage probability in Section 4.3. In Section 4.4, the dependency of system capacity on traffic activity is investigated under the same transmission rate and under the same average rate, and the capacities analyzed with the AILM and the SILM are compared with each other with respect to the traffic activity. Finally, concluding remarks are made in Section 4.5.

4.2 Traffic Modeling

Although traffic characteristics of cellular networks are hard to predict, a number of voice and data models are reported as ON/OFF source models [9, 10] (see Figure 4.2).

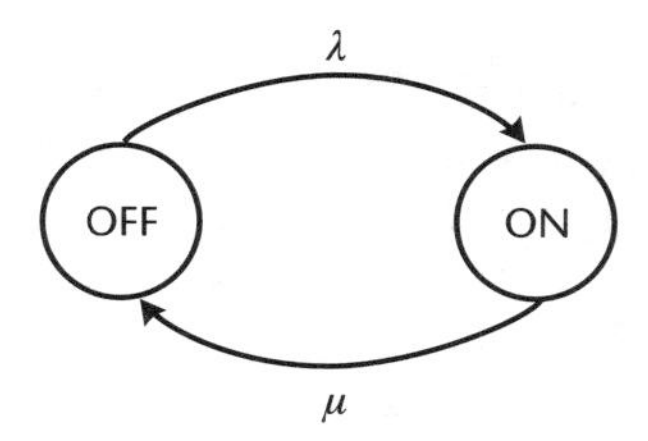

Figure 4.2 ON-OFF source models for voice.

Regarding voice source traffic modeling, it is well known that the process of a voice call transitioning an ON state to an OFF state can be modeled as a two-state Markov chain [9]. The state transition diagram shown in Figure 4.2 depicts how the state transition occurs in such a way that the amount of time spent in each state is exponentially distributed and, given the present state of source traffic, the future is independent of the past. If we assume that the OFF and ON rate from ON to OFF is μ, and from OFF to ON is λ, then the ON period endures for a random time with exponential distribution of parameter λ and then jumps to a silence state with an exponential distribution of parameter μ, and further the average length of the ON and OFF periods is $1/\mu$ and $1/\lambda$, respectively. When a source is ON, it generates packets with a constant interarrival time. When the source OFF, it does not generate any cells.

From the ON/OFF model, the voice activity factor defined as the probability that the state is ON, α, also can be calculated from the balance equations where in this case the activity factor is given as

$$\begin{aligned}\alpha &= \frac{E[ON\ duration]}{E[ON\ duration] + E[OFF\ duration]} \\ &= \frac{1/\mu}{1/\mu + 1/\lambda} = \frac{\lambda}{\lambda + \mu}\end{aligned} \tag{4.1}$$

Acceptable values for $1/\lambda$ and $1/\mu$ for voice calls, the mean ON and OFF times, are 0.35 and 0.65 seconds, respectively. This results in a voice activity factor α of approximately 0.4 [11].

For data traffic such as Web traffic, it has been shown that the probability of large file sizes is not negligible and that the ON duration is effectively characterized by *heavy-tailed* models. The OFF duration is determined by the user's think time, which is also modeled as heavy tailed [12]. A random variable X can be said to have a heavy-tailed distribution if its complementary cumulative distribution function (CDF) has

$$P_r\{X > x\} \sim x^{-\sigma} \tag{4.2}$$

as $x \to \infty$ where $0 < \sigma < 2$. Roughly speaking, the asymptotic shape of the distribution follows a power law, in contrast to exponential decay. Heavy-tailed distribution, by definition, implies that a large portion of the probability mass moves to the tail of distribution as σ decreases.

One of the simplest heavy-tailed distributions is the Pareto distribution, which is power law over its entire range. Along this line, we can assume that the data traffic of each user can be an ON/OFF process where both ON and OFF periods are Pareto distributed such that

$$P_r\{X > x\} = (k / x)^{\sigma} \tag{4.3}$$

where the positive constant k denotes the smallest possible value of the random variable X. Several parameters in ON and OFF processes of real data traffic, for example, are specified in [13]:

- $k_{min,on}$: This is the minimal ON duration, which is determined by the minimum file size and transmission rate. When the minimal file size for Web traffic is about 2k bytes, and CDMA systems provide an average service of about 100 Kbps for each user, then $k_{min,on}$ is about 0.2 second for each burst transmission.
- $k_{min,off}$: This is the minimal OFF duration, which is mainly determined by the user's think time. It varies from about 1 second to 30 seconds. It is reasonable to choose $k_{min,off}$ as 2 seconds.
- σ_{on}: This is determined by the slope of file size distribution, and its typical value is 1.3.
- σ_{off}: This is determined by the slope of think time distribution, and its typical value is 1.5.

Similarly to the case of voice traffic, the activity factor of data traffic α, defined as the probability that the state is ON, can be given as

$$\alpha = \frac{E[ON\ duration]}{E[ON\ duration] + E[OFF\ duration]} \tag{4.4}$$

4.3 Outage Probability and System Capacity

In CDMA systems, although there is no hard limit on the number of concurrent users, there is a practical limit to control the interference between users having the same pilot signal; otherwise, the system can fall into the outage state where QoS requirements cannot be guaranteed. In order to analyze the system capacity of a CDMA system supporting multiclass services in terms of the number of concurrent users with the SILM, and to further investigate the effect of traffic activity on the capacity, the following assumptions are taken:

1. Reverse link is considered.
2. There is perfect power control.
3. Background noise can be neglected.
4. There are N distinct service classes in the system. Each class is characterized by its own QoS requirements composed of the transmission rate and the required bit energy-to-interference power spectral density ratio. Users in the same class have the same QoS requirements.

5. Each user has an ON/OFF traffic activity represented by a binomial random variable such as

$$\phi = \begin{cases} 1, & \text{with probability } \alpha \\ 0, & \text{with probability } 1-\alpha \end{cases} \tag{4.5}$$

where α corresponds to the mean of traffic activity, or the traffic activity factor, which can be calculated based on (4.1) and (4.4) in the case of voice and data, respectively. The traffic activity variables of users in the same class are assumed to be independent and identically distributed (IID), and those of users in different classes are also assumed to be independent.

6. Users transmit information data at a transmission rate in active (ON) state and stop transmitting information data in idle (OFF) state.

When all concurrent users are in active state or the activity factors of the users are equal to one, the number of concurrent users that can be accommodated by the system under nonoutage condition while the QoS requirements of all users are satisfied is limited as [3]

$$\sum_{n=1}^{N} \gamma_n l_n \leq 1 \tag{4.6}$$

where n is the index for service class, l_n is the number of active users in the service class n, and

$$\gamma_n = \left(\frac{W / R_n}{(E_b / I_0)_n} + 1 \right)^{-1} \tag{4.7}$$

W is the allocated frequency bandwidth, R_n is the transmission rate of active users, and $(E_b/I_0)_n$ is the required bit energy-to-interference power spectral density ratio of users in the service class n. In the system supporting N distinct services, the number of active users is defined as a vector, $(l_1, l_2, \ldots, l_N)$ where l_n is an integer for $n = 1, 2, \ldots, N$. Equation (4.6) specifies a capacity plane confining the number of possible active users in the N dimensional space, where it is noteworthy that the capacity per service changes linearly with respect to the capacity variation of the other services. All points $(l_1, l_2, \cdots, l_N)$ under the capacity plane represent acceptable numbers of active users in the system. Total system resources and the resources used by one active user in the service class n correspond to 1 and γ_n, respectively, and γ_n has different values according to the QoS requirements of the class. Equation (4.6) means that the resources used by active users should not exceed total system resources.

4.3.1 AILM

With the traffic activity of users, the number of concurrent users in CDMA systems is confined not generally by the bound for active users in (4.6) but by a looser bound for capacity improvement due to traffic activity. In the AILM, it is assumed that the interference from concurrent users is reduced by the mean of traffic activity (i.e., the random characteristics of traffic activity are simply considered as the mean of traffic activity). By using the assumption, the capacity bound in (4.6) can be modified to [3, 4]

$$\sum_{n=1}^{N} \gamma_n^* k_n \leq 1 \tag{4.8}$$

where k_n is the number of concurrent users in the service class n, and

$$\gamma_n^* = \left(\frac{W / (\alpha \cdot R_n)}{(E_b / I_0)_n} + 1 \right)^{-1} \tag{4.9}$$

Comparing (4.7) with (4.9), it is observed that the instantaneous rate $\phi \cdot R$ in the AILM is assumed to be averaged out such that its average term, $\alpha \cdot R$ is only considered in the capacity analysis, as in Figure 4.3. For the case of AILM, the instantaneous amount of resources used by one user in the system becomes deterministic, and (4.8) becomes a deterministic bound on the number of concurrent users.

4.3.2 SILM

In the SILM, traffic activity is modeled as a binomial random variable to consider capacity influence of traffic activity property. When modeling the activity of concurrent users as binomial random variables, the number of active users l_n in (4.6) becomes a random variable as follows.

$$l_n = \sum_{i=1}^{k_n} \phi_{n(i)} \tag{4.10}$$

where k_n is the number of concurrent users in the service class n and $\phi_{n(i)}$ is a binomial random variable with $P\{\phi_{n(i)} = 1\} = \alpha_n$ representing the traffic activity of the user i in the service class n.

If the number of active users out of concurrent users becomes larger than the bound in (4.6), the outage occurs and QoS requirements of users are not guaranteed. The outage probability can be expressed as

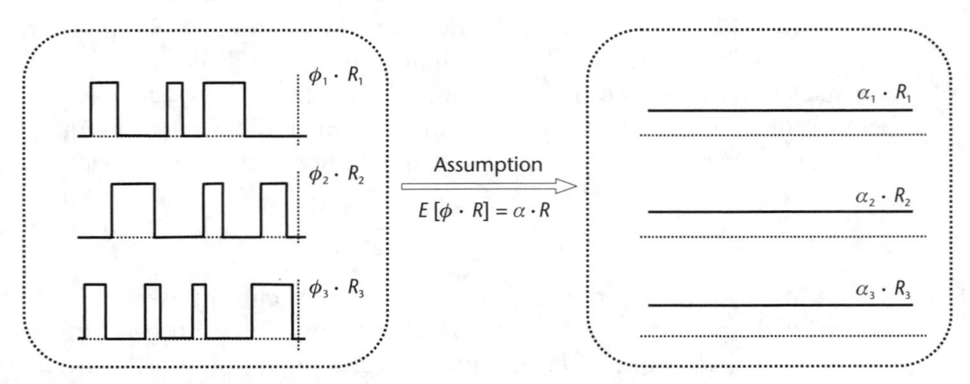

Figure 4.3 Assumption in the AILM.

$$P_0 = \Pr\left\{\sum_{n=1}^{N} \gamma_n \sum_{i=1}^{k_n} \phi_{n(i)} > 1\right\} \tag{4.11}$$

With the assumption that the traffic activity variables of users in the same class are IID, and the traffic activity variables of users in different classes are independent, (4.11) can be modified to

$$P_0 = \sum_{l \notin \Omega} \left\{\prod_{n=1}^{N} P_{k_n}(l_n)\right\} \tag{4.12}$$

where l denotes the number of active users, $(l_1, l_2, \ldots, l_N)$, and $P_{k_n}(l_n)$ represents the probability that the l_n users out of k_n concurrent users are in active state:

$$P_{k_n}(l_n) = \binom{k_n}{l_n} \alpha_n^{l_n} (1-\alpha_n)^{k_n - l_n} \tag{4.13}$$

In (4.12), the Ω represents the set of number of active users in which the outage does not occur, and it can be expressed as

$$\Omega = \left\{l : \gamma \cdot l^T \le 1\right\} \tag{4.14}$$

where $\gamma = (\gamma_1, \gamma_2, \ldots, \gamma_N)$.

For example, let's consider a system supporting two service classes: the service class 1 is for voice service, and the service 2 is for data service. Under the spreading bandwidth $W = 1.25$ MHz and the given transmission rate and required bit energy-to-interference power spectral density ratio in Figure 4.4, the solid line represents the bound on the number of active users under the nonoutage condition.

The set of the number of active users under the bound corresponds to Ω. If the number of active users in the system exceeds the bound, then the outage occurs. For the case that there are 10 and 9 concurrent users in the voice and data service classes, respectively, the number of active users can vary within the rectangular

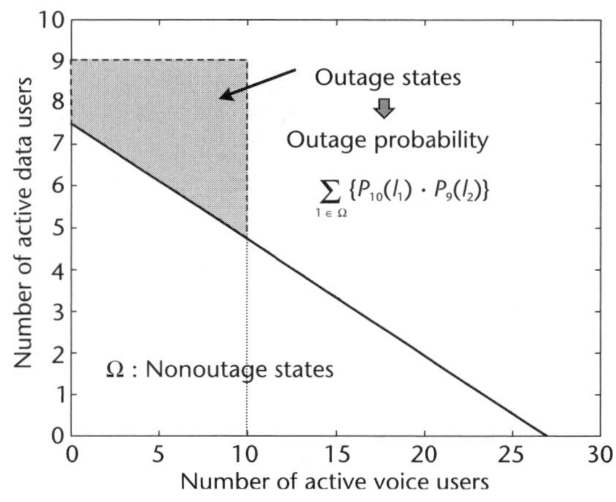

Figure 4.4 Outage and nonoutage sets for $k_v = 10$ and $k_d = 9$ [$R_v = 9.6$ Kbps, $R_d = 19.2$ Kbps, $(E_b/I_0)_v = 7$ dB, and $(E_b/I_0)_d = 10$ dB].

area, and the shadowed area corresponds to the set of the number of active users in which the outage occurs. Then, the outage probability is the sum of the probabilities of all numbers in the shadowed area.

The system capacity is determined under the condition that the outage probability does not exceed the required threshold [1, 5]. The capacity bound on the number of concurrent users can be expressed as

$$\sum_{l \notin \Omega} \left\{ \prod_{n=1}^{N} P_{k_n}(l_n) \right\} \leq P_{0_{req}} \tag{4.15}$$

where $P_{0_{req}}$ is the required outage probability.

Figure 4.5 shows capacity bounds on the number of concurrent users for different values of the required outage probability, where the traffic activity factors for service class 1 and 2 are given as 3/8 and 1/8, respectively. The lower capacity line represents the bound on the number of active users under the nonoutage condition. It also corresponds to the bound on the number of concurrent users under the nonoutage condition because the number of active users can exceed the bound when the number of concurrent users is out of the bound. In Figure 4.5, it is observed that more capacity improvement can be achieved by allowing the outage constraint to be looser. It means that the capacity improvement from the traffic activity is achieved at the expense of the outage probability. By allowing a 1% outage probability, the maximum number of concurrent voice users becomes about twice from 26 to 49, and the maximum number of concurrent data users becomes about three and a half times from 7 to 25. About twofold capacity improvement from voice traffic activity is the same as the result in [1] with the 1% outage probability.

4.3.3 Comparison of AILM and SILM

In the AILM, it is assumed that the traffic activity can be simply considered by the mean value of traffic activity, which implies that the transmission of users is

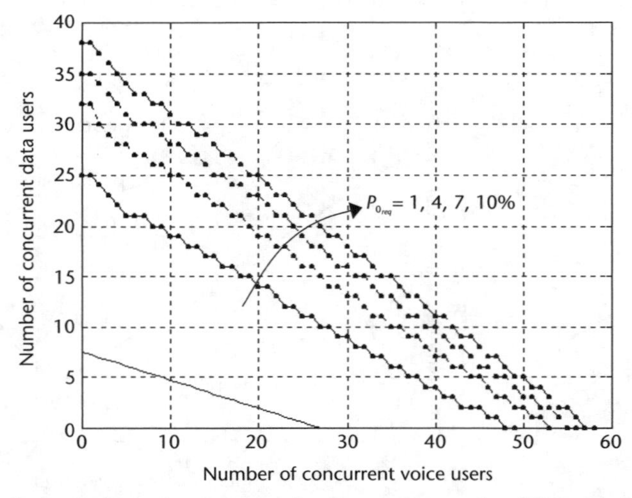

Figure 4.5 Capacity bounds on the number of concurrent users for different values of the required outage probability [R_v = 9.6 Kbps, R_d = 19.2 Kbps, $(E_b/I_0)_v$ = 7 dB, $(E_b/I_0)_d$ = 10 dB, α_v = 3/8, and α_d = 1/8].

regarded as the average rate. From the assumption, interference from concurrent users is reduced by a factor of the mean of traffic activity. However, the outage could occur when the instantaneous interference from concurrent users exceeds the average interference. Because the probability that the instantaneous interference from concurrent users is above the average interference cannot be negligible, the SILM where the traffic activity is modeled as a binomial random variable is a more realistic way to analyze the capacity.

Figure 4.6 shows capacity bounds on the number of concurrent users analyzed with the AILM and the SILM, respectively. The solid line represents the capacity bound with the AILM, and the dotted lines represent the capacity bounds with the SILM for different values of the required outage probability. Comparing the capacity bounds with the AILM and those with the SILM, it is observed that the bound with the AILM places between the bounds with the SILM for 30% and 50% of the required outage probability. It means that the probability that the instantaneous interference from concurrent users exceeds the average interference is about 30–50%. Although the capacity is analyzed under the nonoutage condition with the AILM, 30–50% of the outages are actually due to the simplified assumption on the traffic activity in capacity analysis. The capacity analyzed with the AILM can be said to be more optimistic. Consequently, if the system is operated with the capacity bound analyzed with AILM, then the outages could occur very frequently. Another comparison in terms of the traffic activity will be discussed in the next section.

4.4 Effect of Traffic Activity on System Capacity

In this section, we investigate the effect of traffic activity on system capacity from two points of view. One viewpoint is to analyze the effect of traffic activity under the same transmission rate, and the other is to analyze the effect of traffic activity under the same average rate.

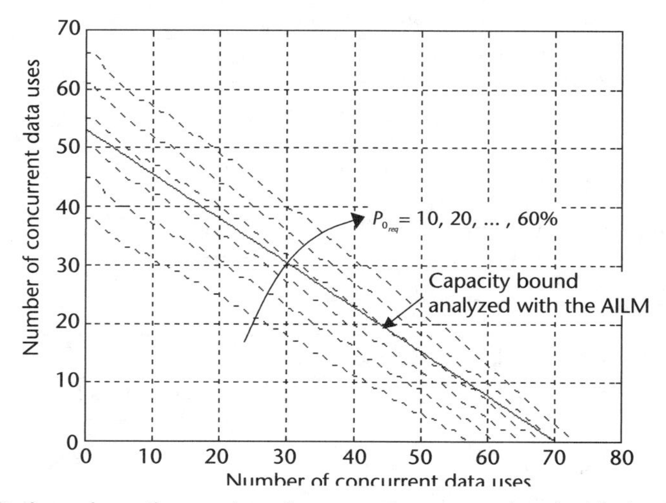

Figure 4.6 Capacity bounds on the number of concurrent users analyzed with the AILM and the SILM. The solid line represents the capacity bound with the AILM, and the dotted lines represent the capacity bounds with the SILM for different values of the required outage probability [R_v = 9.6 Kbps, R_d = 19.2 Kbps, $(E_b/I_0)_v$ = 7dB, $(E_b/I_0)_d$ = 10 dB, α_v = 3/8, and α_d = 1/8].

4.4.1 Analysis Under the Same Transmission Rate

Under the same transmission rate, the average rate is proportional to the activity factor, as shown in Figure 4.7(a). In this case, the average amount of data to be transmitted increases as the activity factor gets increased such that it can be easily expected that the system capacity increases as the activity factor becomes smaller.

A good example of this approach is the capacity improvement with the help of voice activity detection in the CDMA systems supporting voice.

Figure 4.8 shows the maximum number of concurrent users according to the traffic activity factor for the system supporting a single service class with 9.6 Kbps of the transmission rate. In both the AILM and the SILM, it is observed that the maximum number of concurrent users exponentially increases as the traffic activity factor decreases. In the case of AILM, as the interference generated by concurrent users is assumed to be reduced by a factor of the mean of the traffic activity, the capacity improvement is inversely proportional to about α. In particular, the capacity improvement is about 8/3 from 26 to 70 for the voice traffic with $\alpha = 3/8$. However, the net improvement in capacity due to the traffic activity might be smaller than $1/\alpha$ due to the randomness of traffic activity. Subsequently, we can observe that in the case of SILM with a 1% outage probability, the capacity improvement from voice traffic activity is about 2 from 26 to 49.

For a system supporting multiclass services, we consider two service classes: voice and data. Figure 4.9 shows the capacity bounds on the number of concurrent voice and data users, which is analyzed with the SILM for different traffic activity factors of data users under the same transmission rate. As the traffic activity factor

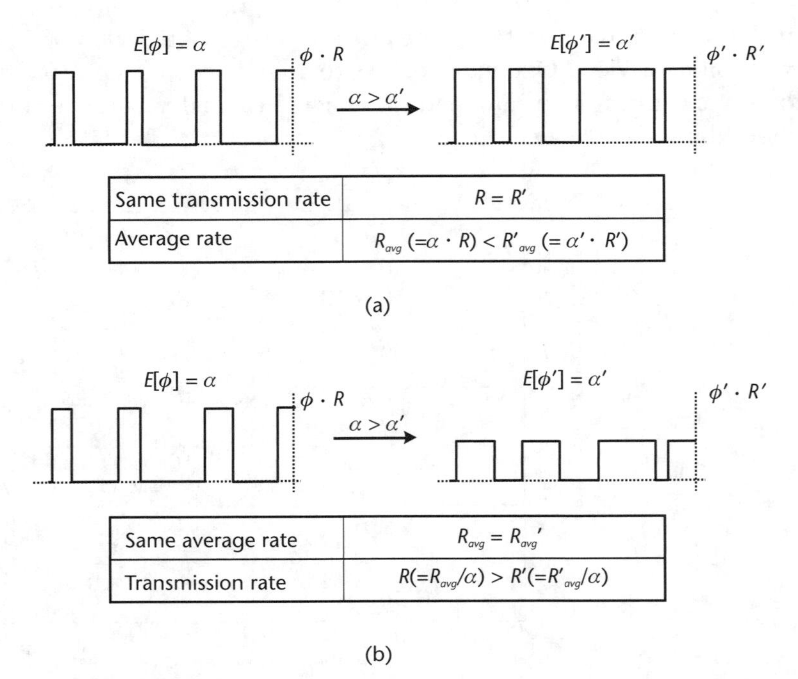

Figure 4.7 Two different viewpoints for investigating the effect of traffic activity on the system capacity: (a) under the same transmission rate, where the average rate R_{avg} changes according to the activity factor; and (b) under the same average rate, where the transmission rate R changes according to the activity factor.

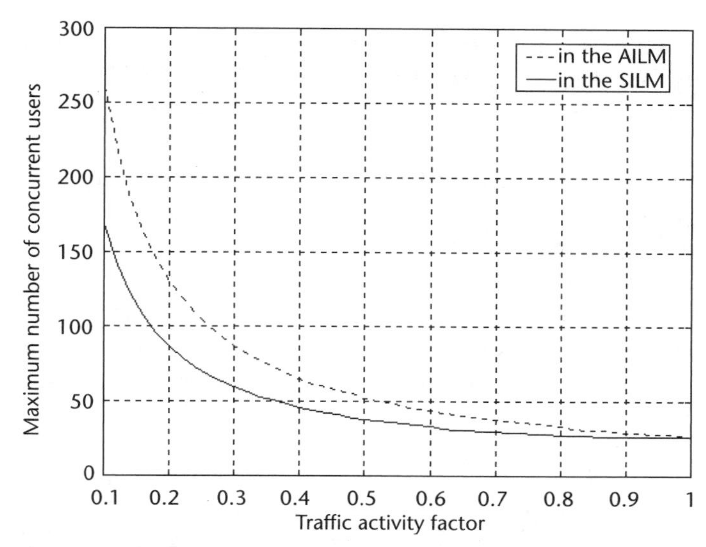

Figure 4.8 Maximum number of concurrent users according to the traffic activity factor under the same transmission rate for the system supporting single service class [R = 9.6 Kbps, (E_b/I_0) = 7 dB, and $P_{0_{req}}$ = 1% in the SILM].

of data users decreases, more voice and data users can be accommodated by the system as with the single service case.

4.4.2 Analysis Under the Same Average Rate

Under the same average rate, the transmission rate changes according to the activity factor, as shown in Figure 4.7(b). The effect of traffic activity on the system capacity under the same average rate is not easily expected because the average amount of information data to be transmitted is the same, regardless of the activity factor.

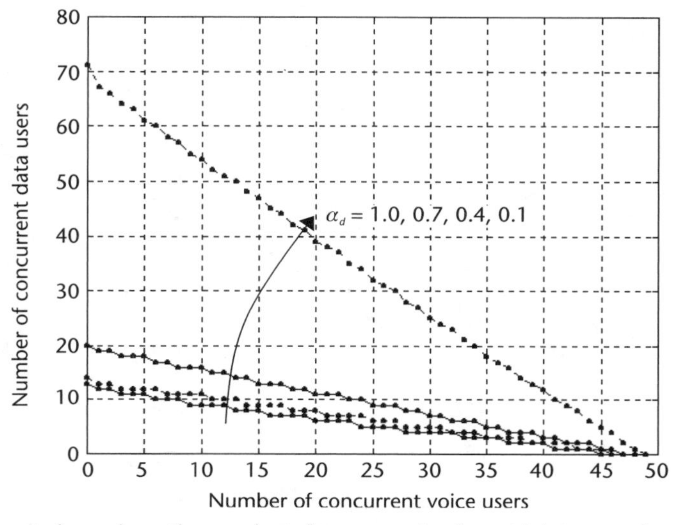

Figure 4.9 Capacity bounds on the number of concurrent voice and data users for different traffic activity factors of data users under the same transmission rate [R_v = 9.6 Kbps, R_d = 10 Kbps, $(E_b/I_0)_v$ = 7 dB, $(E_b/I_0)_d$ = 10 dB, α_v = 3/8, and $P_{0_{req}}$ = 1%].

Figure 4.10 shows the maximum number of concurrent users according to the activity factor for the system supporting a single service class with 3.6 Kbps of the average rate. In the case of the AILM, it is observed that the maximum number of concurrent users is not dependent on the traffic activity factor. It is mainly because the traffics in the AILM have the same average rate and the required bit energy-to-interference power spectral density ratio are treated as the same traffic, regardless of the random characteristics of traffic activity. On the other hand, in the case of the SILM, it is observed that the maximum number of concurrent users tends to increase as the traffic activity factor becomes larger. The partial decrease in the maximum number of concurrent users is caused by the fact that the number of users should be an integer. For the system supporting a single service class, the outage probability in (4.11) becomes

$$P_0 = \Pr\left\{ l = \sum_{i=1}^{k} \phi_i > \left\lfloor \frac{1}{\gamma} \right\rfloor = \left\lfloor \frac{W / R}{E_b / I_0 + 1} + 1 \right\rfloor = \left\lfloor \frac{W / \left(R_{avg} / \alpha\right)}{E_b / I_0 + 1} + 1 \right\rfloor \right\} \tag{4.16}$$

where $\lfloor x \rfloor$ represents the largest integer that is smaller than or equal to x. As the traffic activity factor α becomes larger, $1/\gamma$ linearly increases, but $\lfloor 1 / \gamma \rfloor$ does not change abruptly to the next integer. Until $\lfloor 1 / \gamma \rfloor$ increases to the next integer, the increment of α influences only the random variable l representing the number of active users and increases the outage probability for a certain number of concurrent users, which eventually results in the partial decrease in the maximum number of concurrent users. However, it is noteworthy that the overall effect of the traffic activity under the same average rate results in increased capacity as the traffic activity factor increases. The capacity increment stems from the fact that as the activity factor gets larger under the same average rate, the variance of the interference generated by a certain number of concurrent users decreases, although the average interference is almost the same, which eventually results in a decreased outage probability. In this

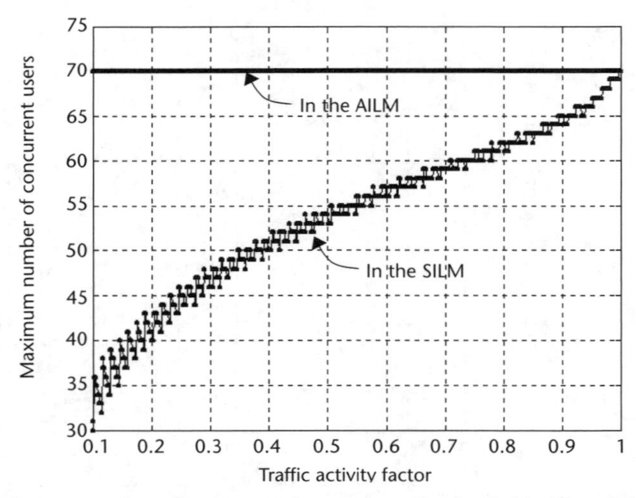

Figure 4.10 Maximum number of concurrent users according to the traffic activity factor under the same average rate for CDMA systems supporting a single service class [R_{avg} = 3.6 Kbps, (E_b/I_0) = 7 dB, and $P_{0_{req}}$ = 1% in the SILM].

case, the system can support about 32 users and 70 users for data traffics with $\alpha \approx$ 0.1 and 1, respectively, while supporting about 50 users for voice traffic with $\alpha \approx$ 3/8. With the observation of the overall effect of traffic activity on the capacity, it can be said that users with smaller activity factors require more system resources in the case of the same average. From Figure 4.10, it can also be observed that the capacity analyzed with the AILM is the same as that analyzed with the SILM when α = 1. It means that the capacity analyzed with AILM corresponds to the upper bound for users with the same average rate.

For a system supporting multiclass services, we consider voice and data service classes. Figure 4.11 shows the capacity bounds on the number of concurrent voice and data users that are analyzed with the SILM for different traffic activity factors of data users under the same average rate. As the traffic activity factor of data users increases under the same average rate, more voice and data users can be accommodated in the system. The capacity bound analyzed with the SILM when α_d = 1.0 in Figure 4.11 is equivalent to the capacity bound analyzed based on the AILM.

Figure 4.12 shows the capacity bounds on the number of concurrent voice and data users for different transmission rates of data users when the traffic activity factor of data users is fixed to 1/8. As expected, the capacity bound decreases as the transmission rate of data users gets larger for a fixed traffic activity factor.

By comparing the capacity bounds analyzed with the AILM with those with the SILM, we can observe that the capacity bound based on the SILM gets closer to the capacity bound based on the AILM as the activity factors of both voice and data users increase under the same average rate.

4.5 Conclusions

In this chapter, the capacity of a CDMA system supporting multiclass services with ON/OFF traffic activity has been investigated by modeling the traffic activity as a

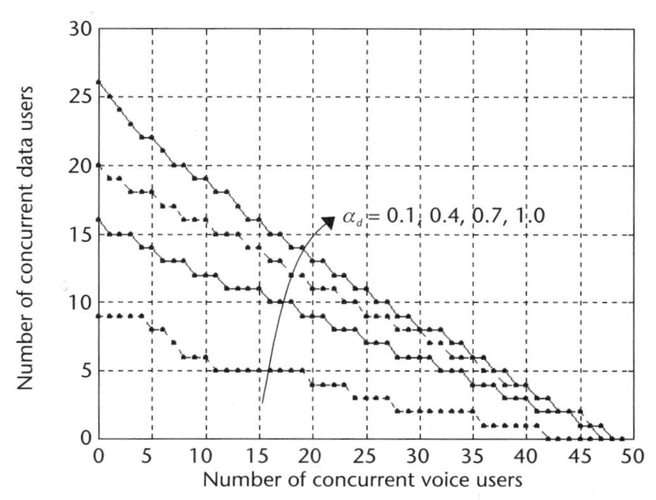

Figure 4.11 Capacity bounds on the number of concurrent voice and data users for different traffic activity factors of data users under the same average rate in the SILM [R_v = 9.6 Kbps, $R_{d_{avg}}$ = 5 Kbps, $(E_b/I_0)_v$ = 7 dB, $(E_b/I_0)_d$ = 10 dB, and $P_{0_{req}}$ = 1%].

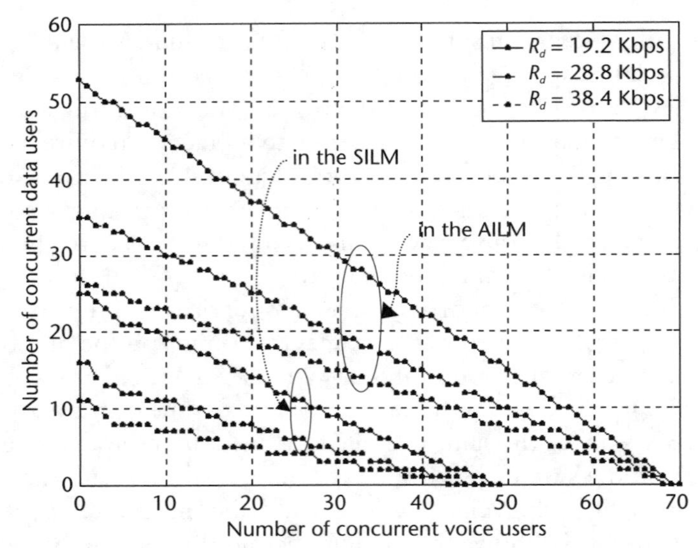

Figure 4.12 Capacity bounds on the number of concurrent voice and data users for different transmission rates of data users when the traffic activity factor of data users is fixed to 1/8.

binomial random variable, a method called the SILM. The corresponding capacity according to the outage probability shows that the capacity improvement due to the traffic activity is achieved at the cost of the outage. It is also observed that tens of percent of outage could occur practically if system capacity is analyzed by simplifying the traffic activity just as its mean value, a method known as the AILM. The effect of traffic activity on the system capacity under the same transmission rate and under the same average rate has also been investigated. As the traffic activity factor gets larger, the system capacity increases under the same average rate, while it decreases under the same transmission rate. In the case of the same average rate, it is also observed that users with smaller traffic activity factors make use of more system resources, although the users transmit the same amount of information data in a certain duration. The capacity analyzed by the AILM is able to represent the trend of capacity variation according to the traffic activity factor under the same transmission rate. However, the AILM is unable to represent the trend of capacity variation under the same average rate, and the capacity analyzed by the AILM corresponds to the upper bound of the capacity for the traffic with the same average rate.

References

[1] Gilhousen, K. S., et al., "On the Capacity of a Cellular CDMA System," *IEEE Trans. on Vehicular Technology,* 1991, pp. 303–312.

[2] Timotijevic, T., and J. A. Schormans, "ATM-Level Performance Analysis on a DS-CDMA Satellite Link Using DTX," *IEE Proceedings—Communications*, 2000, pp. 47–56.

[3] Yang, Y. R., et al., "Capacity Plane of CDMA System for Multimedia Traffic," *IEE Electronics Letters,* 1997, pp. 1432–1433.

[4] Kim, K., and Y. Han, "A Call Admission Control Scheme for Multi-Rate Traffic Based on Total Received Power," *IEICE Trans. on Communications*, 2001, pp. 457–463.

[5] Matragi, W., and S. Nanda, "Capacity Analysis of an Integrated Voice and Data CDMA System," *IEEE Proc. of Vehicular Technology Conference,* 1999, pp. 1658–1663.

[6] Viterbi, A. M., and A. J. Viterbi, "Erlang Capacity of a Power-Controlled CDMA System," *IEEE Journal on Selected Areas in Communications*, 1993, pp. 892–900.

[7] Ayyagari, D., and A. Ephremides, "Cellular Multicode CDMA Capacity for Integrated (Voice and Data) Services," *IEEE Journal on Selected Areas in Communications*, 1999, pp. 928–938.

[8] Kim, D. K., and D. K. Sung, "Capacity Estimation for an SIR-Based Power-Controlled CDMA System Supporting On-Off Traffic," *IEEE Trans. on Vehicular Technology*, 2000, pp. 1094–1101.

[9] Brady, P. T., "A Statistical Analysis of On-Off Patterns in 16 Conversations," *Bell System Technical Journals*, 1968, pp. 73–91.

[10] Willinger, W., et al., "Self-Similarity Through High-Variability: Statistical Analysis of Ethernet LAN Traffic at the Source Level," *IEEE/ACM Trans. on Networking*, Vol. 5, 1997, pp. 71–86.

[11] Sriram, K., and W. Whitt, "Characterizing Superposition Arrival Processes in Packet Multiplexers for Voice and Data," *IEEE Journal on Selected Areas in Communications*, 1986, pp. 833–846.

[12] Park, K., and W. Willinger, *Self-Similar Network Traffic and Performance Evaluation*, New York: John Wiley & Sons, 2000.

[13] Zang, J., M. Hu, and N. Shroff, "Bursty Data over CDMA: MAI Self Similarity, Rate Control and Admission Control," *IEEE Proc. of Infocom*, 2002, pp. 391–399.

CHAPTER 5

A Dynamic Resource Allocation Scheme to Efficiently Utilize System Capacity

With today's growing demands for multimedia services and high degree of user mobility, RRM plays a important role in future CDMA systems to efficiently utilize limited radio resources and to provide more mobile users with guaranteed QoS anywhere at any time.

Further, the performance of a system with given physical resource (e.g., given bandwidth of radio spectrum) heavily depends on RRM scheme. Even though the effectiveness and efficiency of the RRM are affected by system characteristics at the physical, link, and network layers, the major objective of RRM is to enhance the capacity (i.e., the maximum number of users or the throughput that can be supported in a given band for a given QoS).

Major radio resource management schemes can be divided into CAC and resource allocation for ongoing calls [1–3].

1. CAC involves the control of both new calls and handoff calls. A new call is a call that originates within a cell and that requests access to the cellular system. A handoff call is a call that originated in one cell but requires and requests resources in another cell. At the network layer, CAC can decide whether a new or handoff connection should be admitted into the system. Admitting more connections than the capacity of lower layers can handle will result in network congestion and the inability to guarantee QoS performance. On the other hand, admitting fewer connections than the capacity of lower layers will underutilize the system resources.
2. Resource allocation for ongoing calls is the distribution of the radio resources among existing users so that the system objective function (e.g., the throughput) can be maximized while maintaining the target QoS (e.g., good voice quality). Power distribution and rate allocation are the basis to achieve this objective in CDMA systems, where system resources are shared by all active users.

The RRM in a voice-centric cellular system is relatively simple. A voice call is admitted if there are any free channels, and speech quality is maintained by preserving a predetermined SIR through power control and handoff. However, emerging next generation cellular systems aim to service both voice users and data users. The RRM in such systems is complex and must be designed carefully.

This book addresses RRM in CDMA systems supporting multiclass services from two perspectives. First, this chapter describes a resource allocation scheme with which we can find the optimum set of data rates for concurrent users and further maximize the system throughput while satisfying the minimum QoS requirements of each user for ongoing connected calls. Second, Chapter 6 presents a CAC scheme for CDMA systems supporting voice and data services to accommodate more traffic load in the system, where some system resources are reserved exclusively for handoff calls to have higher priority over new calls, and additionally queuing is allowed for both new and handoff data traffic not sensitive to delay.

5.1 Introduction

Because wireless systems have limited system resources, and multimedia services have various QoS requirements, the system resources must be carefully managed to achieve high efficiency. In order to fully utilize the system resources of multimedia CDMA systems, we should identify the system capacity, which can be evaluated in terms of the number of concurrent users with various kinds of traffic [4, 5].

In most cases, the system is not fully loaded, and some extra remaining resources exist. For the efficient use of system resources, resource allocation methods should be properly designed to allocate remaining resources to current users in the system for better performance. In the case of CDMA systems, such remaining system resources can correspond to the power or data rate because the capacity of CDMA systems is interference limited. As a study to utilize remaining resources efficiently in CDMA systems, Ramakrishna et al. proposed an efficient resource allocation scheme with the objective of maximizing the throughput for dual traffic case: CBR traffic and VBR traffic [6]. It is also shown that the throughput can be improved by allocating the remaining resources to a limited number of VBR users rather than all VBR users, and eventually it can be maximized when the remaining resources are allocated to one VBR user. However, Ramakrishna et al. considered only the case where there are only single VBR and CBR service groups, where users in one group have the same QoS requirements. In multimedia environments, multiple VBR and CBR groups should be considered in the resource management of CDMA systems. In this chapter, as an expended work of [6], we consider multiple service groups of CBR and VBR traffic in order to include more generalized cases where users in each VBR service group demands different BERs and minimum transmission rate requirements while users in each CBR service group requires distinct BERs and constant transmission rates. Further, we present a dynamic resource allocation scheme with which we can maximize system throughput while satisfying QoS requirements of all VBR and CBR users.

This chapter is organized as follows: In Section 5.2, we review the capacity of multimedia CDMA systems under the system model being considered in order to quantify system resources. In Section 5.3, the system throughput is defined, and the throughput maximization problem is addressed and formulated. With the observations of the previous section, in Section 5.4 we propose the dynamic resource allocation scheme that maximizes the system throughput while satisfying all QoS requirements of users. In Section 5.5, we present some case studies for the operation of the proposed scheme. Finally, we draw some conclusions in Section 5.6.

5.2 System Capacity and Remaining Resources

We consider the reverse link of a CDMA system under perfect power control assumption. For various services in the system, it is assumed that there are M CBR service groups and N VBR service groups. Users in one group have the same QoS requirements. The QoS requirements of a CBR service group are composed of a BER and a constant transmission rate, while those of a VBR service group are composed of a BER and a minimum transmission rate. It is also assumed that the BER requirement can be mapped into an equivalent E_b/I_0 requirement.

In order to satisfy the QoS requirements for all concurrent users, the capacity of the CDMA system is limited as [5]

$$\sum_{i=1}^{M}\gamma_{c_i}k_{c_i} + \sum_{j=1}^{N}\gamma_{v_j}k_{v_j} \leq 1 \tag{5.1}$$

where

$$\gamma_{c_i} = \left(\frac{W}{R_{c_i}q_{c_i}}+1\right)^{-1} \text{ and } \gamma_{v_j} = \left(\frac{W}{R_{v_j}q_{v_j}}+1\right)^{-1} \tag{5.2}$$

k_{c_i} and k_{v_j} denote the number of users in the ith CBR service group ($i = 1, \ldots, M$) and the jth VBR service group ($j = 1, \ldots, N$), respectively. W is the spreading bandwidth. q_{c_i} and q_{v_j} are the required bit energy-to-interference spectral density ratio of the ith CBR service group and the jth VBR group, respectively. R_{c_i} and R_{v_j} are the transmission rate of the ith CBR service group and jth VBR service group, respectively. R_{c_i} is a constant rate while R_{v_j} is a variable rate and should be greater than the required minimum transmission rate.

If $R_{v_j,\min}$ is defined as the required minimum transmission rate, and a_j of the jth VBR group is defined as a rate factor to control the transmission rate of the jth VBR group, then the transmission rate, R_{v_j} can be expressed as

$$R_{v_j} = a_j R_{v_j,\min} \tag{5.3}$$

where $a_j \geq 1$. In (5.1), γ_{c_i} and γ_{v_j} correspond to the amount of system resources used by one user in the ith CBR group and that used by one user in the jth VBR group, respectively. Equation (5.1) means that the system resources used by concurrent users should not exceed total system resources.

The system capacity in (5.1) can be regarded as a bound confining the number of supportable concurrent users. In order to reach the maximum bound of the number of supportable concurrent users, the resources being utilized by a user should be the minimum amount that is needed to satisfy QoS requirements. It can be simply achieved by setting the rate factor, $a_j = 1$ for all VBR groups because CBR groups use a fixed resource. The bound on the number of concurrent users is a hyperplane in $(M + N)$ dimensional space. All points $(k_{c1}, k_{c2}, \ldots, k_{cM}, k_{v1}, k_{v2}, \ldots, k_{vN})$ under the hyperplane represent the possible number of users.

In most cases, the system is not always fully loaded, which implies that there may exist some remaining resources from time to time. Subsequently, efficient resource allocation schemes are needed to readjust system resource and further to utilize the remaining resources more efficiently. The remaining resources can be defined as

$$\Gamma = C - \sum_{i=1}^{M} \gamma_{c_i} k_{c_i} - \sum_{j=1}^{N} \gamma_{v_{j,\min}} k_{v_j} \tag{5.4}$$

where C represents total amount of system resources, and its maximum value is 1, and $\gamma_{v_{j,\min}}$ is given as γ_{v_j} when $a_j = 1$ for $j = 1, \ldots, N$. Under the multiple cell environment, we can consider the effect of intercell interference on the remaining resources by adjusting the value of C.

5.3 Service Rates for Throughput Maximization

One way to efficiently utilize the remaining resources is to allocate the remaining resources to VBR users for the improvement of the throughput. Intuitively, we can increase the transmission rate of VBR users until the remaining resources are exhausted. However, it is a remaining question how to allocate the remaining resources to multiple VBR users. Before approaching the problem, we must consider the relation between the transmission rate and the allocated resources.

Figure 5.1 shows the first derivative of the transmission rate with respect to the allocated resources, $\frac{\partial R}{\partial \gamma}$ for different values of required E_b/I_0, q, which is expressed as follows:

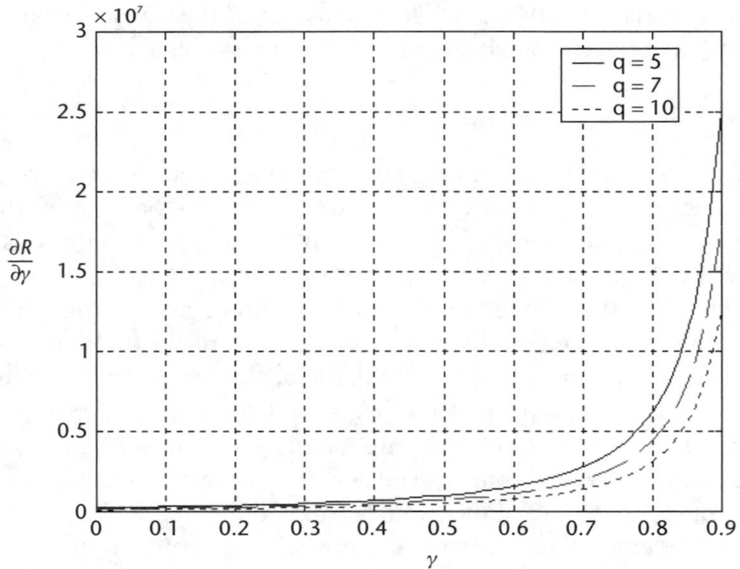

Figure 5.1 First derivative of the transmission rate with respect to the allocated resources, $\frac{\partial R}{\partial \gamma}$ for different values of the required E_b/I_0, q.

$$\frac{\partial R}{\partial \gamma} = \frac{W}{q} \cdot \frac{1}{(1-\gamma)^2} \tag{5.5}$$

It is noteworthy that the transmission rate R exponentially increases as the allocated resources r increases, and the increment rate of R gets larger as the required E_b/I_0, q decreases. The transmission rate does not have a linear relation with the allocated resources. Therefore, it can be expected that the achievable throughput depends on the way of allocating the remaining resources to VBR users.

Figure 5.2 shows an example of system resource status with three service groups. In the figure, three shadowed areas show the resources being utilized by users in one CBR service group and two VBR service groups, respectively, where the resource utilized by one user is the minimum amount needed to satisfy QoS requirements. The blank area represents the remaining resources of the system, and it can be allocated to VBR users in several ways.

Figure 5.3 illustrates an example of allocating the remaining resources to users. Figure 5.3(a) shows the impartial allocation of the remaining resources to all VBR users in the system, while Figure 5.3(b, c) represents the allocation of the remaining resources to users in only a certain VBR service group. The throughput, which can be obtained from each allocation in Figure 5.3(a–c), could be different from one another due to the nonlinear relation between the transmission rate and the allocated resources.

In order to maximize the throughput by allocating remaining resources to multiple VBR users properly, we need to find the optimum transmission rate set for VBR service groups, which corresponds to the optimum resource allocation. Noting

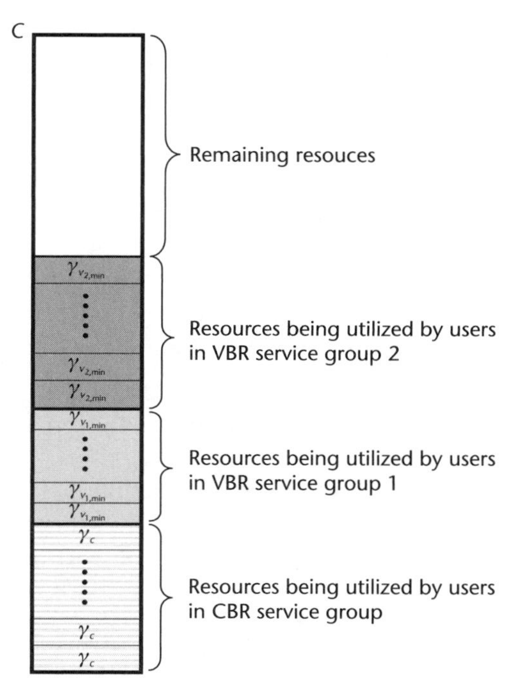

Figure 5.2 System resource status for three service groups: one is a CBR group, and the others are VBR groups.

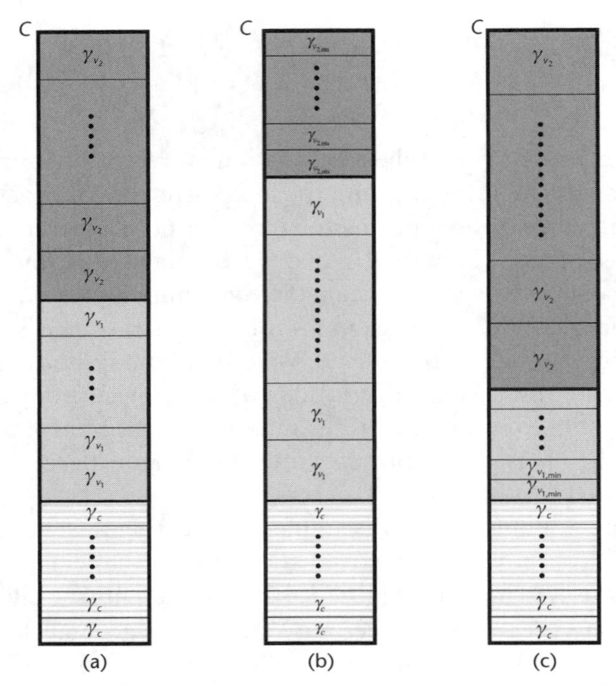

Figure 5.3 Allocation of the remaining resources: (a) shared by all VBR users, (b) shared by users in VBR service group 1, and (c) shared by users in VBR service group 2.

that the transmission rate of VBR service groups is represented by the rate factor, the throughput maximization problem can be formulated as follows:

Find the optimum set $(a_1, a_2, \ldots, a_N)_{opt}$ that maximizes the throughput T,

$$T = \sum_{i=1}^{M} R_{c_i} k_{c_i} + \sum_{j=1}^{N} a_j R_{v_j,\min} k_{v_j} \tag{5.6}$$

subject to

$$\sum_{i=1}^{M} \gamma_{c_i} k_{c_i} + \sum_{j=1}^{N} \gamma_{v_j} k_{v_j} \leq C \tag{5.7}$$

$$a_j \geq 1 \text{ for all } j \tag{5.8}$$

Here we have set $C = 1$ by ignoring the effect of intercell interference on the system.

Under the constraint in (5.7) on the rate factors, it can be shown that the shape of T in (5.6) has a convex form as a function of the rate factors, which can be easily proved by taking the second derivative of T with respect to the rate factor, a_l as follows:

$$\frac{\partial^2 T}{\partial a_l^2} = 2W^2 \frac{Q_{v_l}^2}{Q_{v_N}} \frac{k_{v_N}^2 k_{v_l} R_{v_{N,\min}}}{\left(W + a_l Q_{v_l}\right)^3} \cdot \frac{\left(k_{v_N} - C\right) + \sum_{i=1}^{M} \gamma_{c_i} k_{c_i} + \sum_{j=1}^{N-1} \gamma_{v_j} k_{v_j} + k_{v_l} W / \left(W + a_l Q_{v_l}\right)}{\left(\left(k_{v_N} - C\right) + \sum_{i=1}^{M} \gamma_{c_i} k_{c_i} \sum_{j=1}^{N} \gamma_{v_j} k_{v_j}\right)^3} \tag{5.9}$$

where $Q_{vl} = R_{v_{l,\min}} q_{v_l}$. Noting that $k_{v_N} \geq 1$ and all other terms in (5.9) have a positive value, we know $\partial^2 T / \partial a_j^2 > 0$ for all j. As the constraint in (5.7) confines a feasible region of a_j, a_j has a value between 1 and a_j^{vertex}, where a_j^{vertex} is limited by (5.7) and further occurs when $a_i = 1$ for all $i \neq j$. As T takes a part of the convex shape as shown in (5.9), then the maximum of T occurs at either $a_j = 1$ or $a_j = a_j^{vertex}$. Noting that T has minimum value when $a_j = 1$ for all j, although this is obviously much smaller throughput in the sense that $\sum_{i=1}^{M} \gamma_{c_i} k_{c_i} + \sum_{j=1}^{N} \gamma_{v_{j,\min}} k_{v_j} < C$, we can get the maximum throughput at $a_j = a_j^{vertex}$. It is also noteworthy that the $a_j = a_j^{vertex}$ was derived from (5.7) and $a_j = a_j^{vertex}$ means that $a_i = 1$ for all $i \neq j$. So, we need to test only vertices points to find the maximum throughput.

To visualize these facts, we consider a system with four service groups composed of one CBR group and three VBR groups with distinct QoS requirements. More specific parameters are summarized in Table 5.1. Figure 5.4 shows the relation between the rate factors based on (5.7) and (5.8) when system resources are fully utilized (i.e., $C = 1$). All points (a_1, a_2, a_3) on the surface represent the set of possible rate factors to improve the throughput by using the remaining resources. Figure 5.5 shows the corresponding throughput T according to the set of rate factors of Figure 5.4, where we omit a_3 because a_3 is determined by a_1 and a_2. As previously pointed out, Figure 5.5 shows that T takes a convex shape and $T_{\max}$ is obtained at one of the vertices. In this case, the optimum rate factor set $\hat{a}_{opt}$ is (5.033, 1, 1) with which the throughout T is maximized to 280 Kbps.

5.4 The Proposed Resource Allocation Scheme

From the fact that the maximum throughput $T_{\max}$ can be obtained at one of the vertices, we can reduce the infinite number of the rate factor sets to N candidate sets.

Table 5.1 Parameters of a CDMA System for One CBR Service Group and Three VBR Service Groups

Parameters	*Symbol*	*Value*
Bandwidth	W	1.25 MHz
Constant transmission rate for CBR group	R_c	9.6 Kbps
Minimum transmission rate for the jth VBR service group	$R_{v_{j,\min}}$	9.6, 4.8, 2.4 Kbps for j = 1, 2, 3
Required bit energy-to-interference spectral density ratio for CBR group	q_c	5
Required bit energy-to-interference spectral density ratio for the jth VBR service group	q_{v_j}	5, 7, 10 for j = 1, 2, 3
Number of concurrent users in CBR service group	k_c	2
Number of concurrent users in the jth VBR service group	k_{v_j}	5, 3, 2 for j = 1, 2, 3

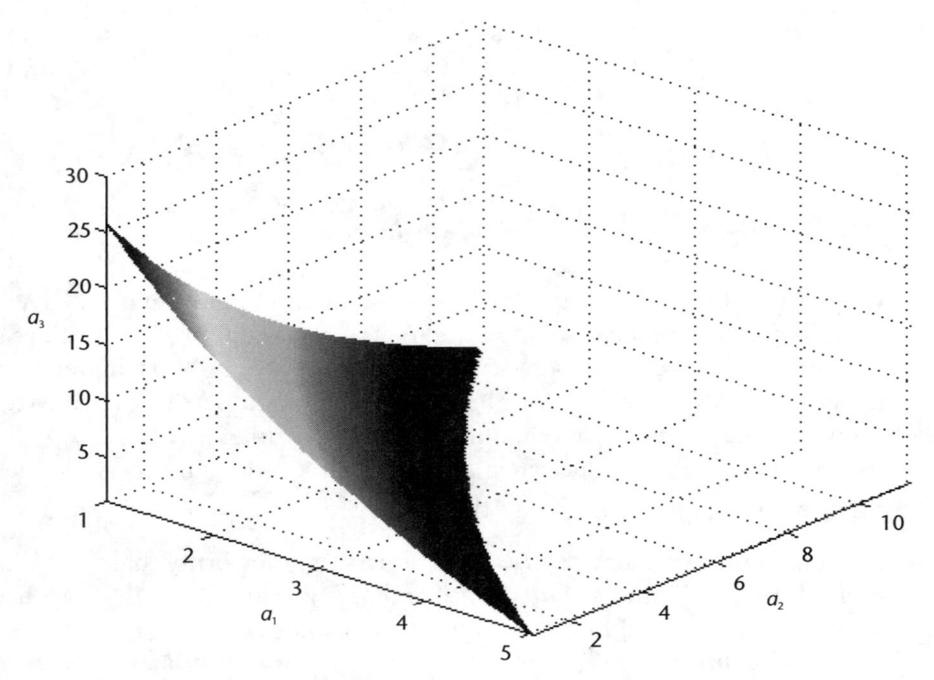

Figure 5.4 Relation between the rate factors for three VBR service groups.

Further, the optimum transmission rate set for the system with M CBR groups and N VBR groups can be found through the following general procedure:

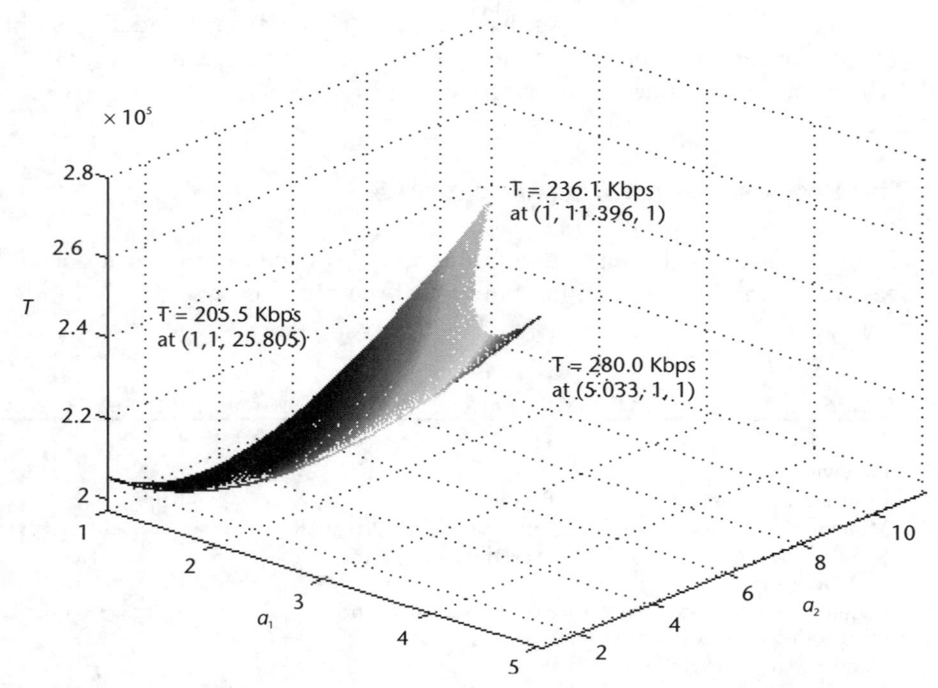

Figure 5.5 Throughput of an example according to the set of rate factors of Figure 5.4.

- Step 1: Calculate N candidate rate factor sets.

$$\overline{A} = \begin{bmatrix} A_1 & 1 & 1 & \cdots & 1 & 1 \\ 1 & A_2 & 1 & \cdots & 1 & 1 \\ 1 & 1 & A_3 & \cdots & 1 & 1 \\ \vdots & \vdots & \vdots & \ddots & \vdots & \vdots \\ 1 & 1 & 1 & \cdots & A_{N-1} & 1 \\ 1 & 1 & 1 & \cdots & 1 & A_N \end{bmatrix} \tag{5.10}$$

where the element of matrix $\overline{A}$, A_k (for $k = 1, \ldots, N$) is calculated with following equation.

$$A_k : \left(\sum_{i=1}^{M} \gamma_{c_i} k_{c_i} + \sum_{j=1}^{N} \gamma_{v_j} k_{v_j} \right) \Bigg|_{a_j = \begin{cases} 1, & j \neq k \\ A_j, & j = k \end{cases}} = C \tag{5.11}$$

In the matrix $\overline{A}$, the nth row vector corresponds to the nth candidate rate factor set.

- Step 2: Calculate the throughput for VBR groups generated by N candidate rate factors.

$$\overline{T}_v = \overline{A} \cdot \overline{R}_{v_{\min}} \tag{5.12}$$

where $\overline{T}_v^t = \left[T_{v_1}, T_{v_2}, \ldots, T_{v_N} \right]$ and $\overline{R}_{v_{\min}}^t = \left[R_{v_{1,\min}}, R_{v_{2,\min}}, \ldots, R_{v_{N,\min}} \right]$

- Step 3: Select the rate factor set generating the maximum throughput.

$$\overline{A}_l = [1, \ldots, 1, A_l, 1, \ldots, 1] \tag{5.13}$$

where

$$l = \arg\max_j \{T_j\}, \text{ for } j = 1, \ldots, N \tag{5.14}$$

- Step 4: Determine the transmission rate set for N VBR groups.

$$\overline{R}_v = \overline{A}_l \cdot \overline{R}_{v_{\min}} \tag{5.15}$$

where $\overline{R}_v^t = \left[R_{v_1}, R_{v_2}, \ldots, R_{v_N} \right]$.

In the proposed procedure, the main objective is to select VBR service groups to have nontrivial transmission rates. It can be simplified by considering the amount of throughput increments for each VBR service group. By allocating the remaining resources to the jth VBR service group, we obtain corresponding throughput increments as follows:

$$\begin{aligned}
\Delta T_{v_j} &= T_{v_j}\Big|_{\gamma_{v_j}=\gamma_{v_j,\min}+\frac{\Gamma}{k_{v_j}}} - T_{v_j}\Big|_{\gamma_{v_j}=\gamma_{v_j,\min}} \\
&= k_{v_j}\frac{W}{q_{v_j}}\left[\frac{\gamma_{v_j,\min}+\frac{\Gamma}{k_{v_j}}}{1-\left(\gamma_{v_j,\min}+\frac{\Gamma}{k_{v_j}}\right)} - \frac{\gamma_{v_j,\min}}{1-\gamma_{v_j,\min}}\right] \\
&= \frac{W}{q_{v_j}}\frac{\Gamma}{\left(1-\gamma_{v_j,\min}-\frac{\Gamma}{k_{v_j}}\right)\left(1-\gamma_{v_j,\min}\right)}
\end{aligned} \tag{5.16}$$

Noting that the constant term $W \cdot \Gamma$ in (5.16) is of no consequence in selecting the group and it can be ignored, we can simplify the group selection as follows:

Select the service group $\hat{j}$ which satisfies

$$\hat{j} = \arg\min_{j}\left\{q_{v_j}\left(1-\gamma_{v_j,\min}-\frac{\Gamma}{k_{v_j}}\right)\left(1-\gamma_{v_j,\min}\right)\right\} \tag{5.17}$$

With this simplified group selection rule, we propose a simple scheme dynamically allocating the remaining resources according to the change of the number of concurrent users. Figure 5.6 shows the overall flow chart of the proposed dynamic resource allocation scheme. In this scheme, the remaining resources are reallocated when a new call is accepted or a call is completed. When a new call attempt is generated, it is determined whether the call is accepted or blocked by comparing the minimum resources required by the user with the remaining resources. If the call is blocked, the reallocation of the remaining resources is not needed because there is no change in the system status. When a call is accepted or completed, the user number set and the remaining resources are updated.

5.5 Group Selection According to the Parameters of VBR Service Groups

In this section, we investigate the trends of group selection of the proposed resource allocation scheme in a system supporting two VBR service groups and the corresponding throughput variations according to the change of the parameters of VBR service groups, such as k_v, $R_{v_{\min}}$, and q_v. VBR service groups are assumed to be distinct from each other in the sense of the parameters k_v, $R_{v_{\min}}$, and q_v.

Figure 5.7 shows the group selection and the contour of corresponding throughput according to the difference in the number of concurrent users, k_v. In this case, the same values of $R_{v_{\min}}$ and q_v are used for both service groups in order to investigate the group selection trends according to k_v. The capacity bound in the figure confines the number of concurrent users that can be supportable in the system.

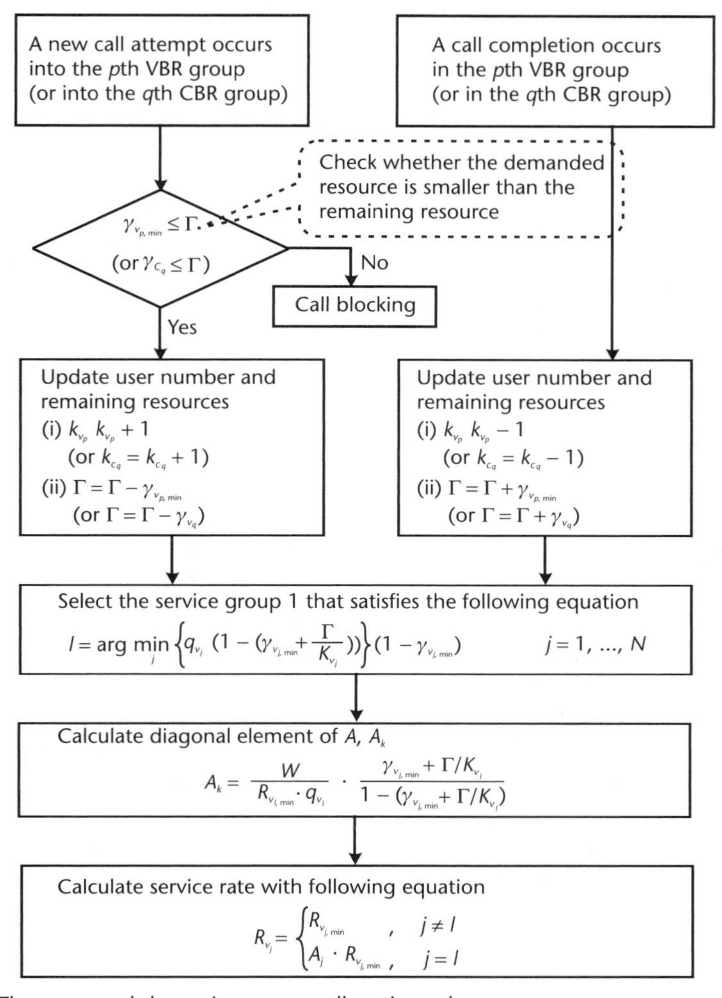

Figure 5.6 The proposed dynamic resource allocation scheme.

From Figure 5.7, it is observed that the group with smaller number of concurrent users is selected. For example, if there are eight users in service group 1 and two users in service group 2 in the system, then eight users in service group 1 will get the minimum transmission rate and the remaining resources will be fully allocated to two users in the service group 2 to maximize the system throughput. Figure 5.7 also shows that the smaller the number of concurrent users in the system, the more throughput is achieved. Further, we know that the more uneven the number of concurrent users in the service group, the more throughput is obtained, which is the same result as in [6].

Figure 5.8 shows the group selection and the contour of corresponding throughput according to the difference in the required minimum transmission rate, $R_{v_{\min}}$. In this case, the same values of k_v and q_v are set for both service groups in order to observe group selection trends according to $R_{v_{\min}}$. The capacity bound in the figure confines the maximum value of $R_{v_{\min}}$ that can be allowable in the system with given k_v and q_v. From Figure 5.8, it is observed that the service group with the larger value

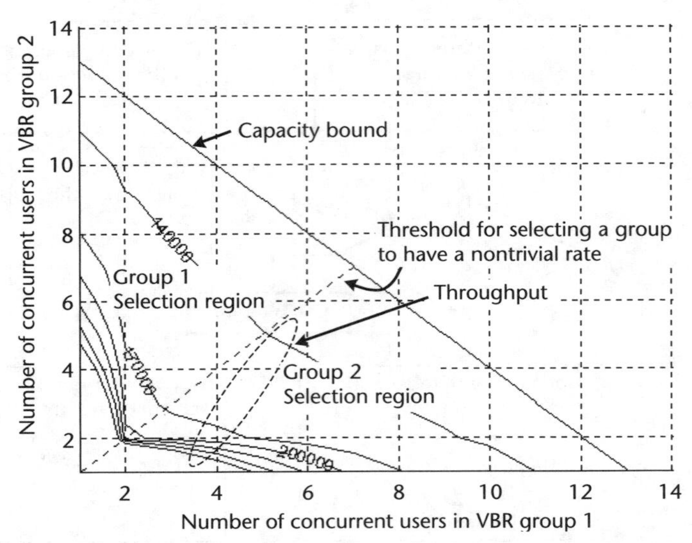

Figure 5.7 Group selection trends and the contour of corresponding throughput according to the difference in the number of concurrent users when q_{v_1} and q_{v_2} are 10, and $R_{v_{1\min}}$ and $R_{v_{2\min}}$ are 9.6 Kbps.

of $R_{v_{\min}}$ is selected; further, as the value of $R_{v_{\min}}$ get smaller, we can get more throughput. However, the minimum transmission rate of one group does not have an influence on the throughput in the selection region of the group for the fixed minimum transmission rate of the other group. The reason it has no influence is that the increment of the rate factor by using the remaining resources is equivalent to the increment of the minimum transmission rate. Therefore, the transmission rate supported by the system for one group is the same as another, irrespective of the minimum transmission rate in the selection region of the group.

Figure 5.9 shows the group selection and the contour of corresponding throughput according to the difference in the required bit energy-to-interference spectral

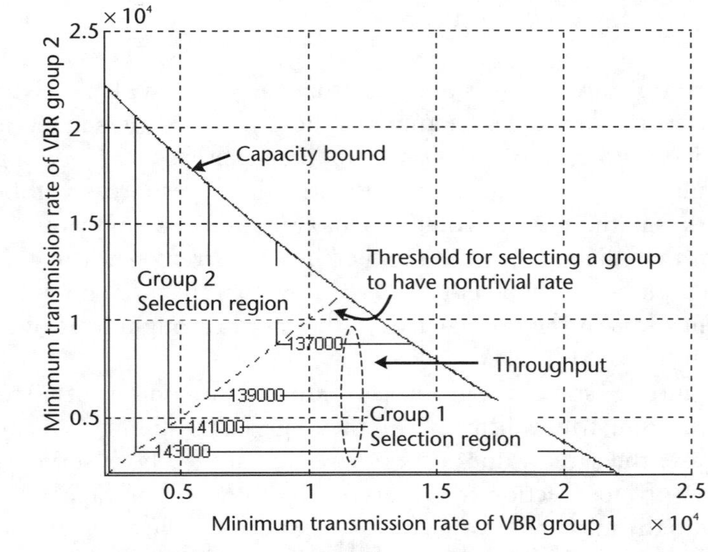

Figure 5.8 Group selection trends and the contour of corresponding throughput according to the difference in the minimum transmission rate when q_{v_1} and q_{v_2} are 10, and k_{v_1} and k_{v_2} are 6.

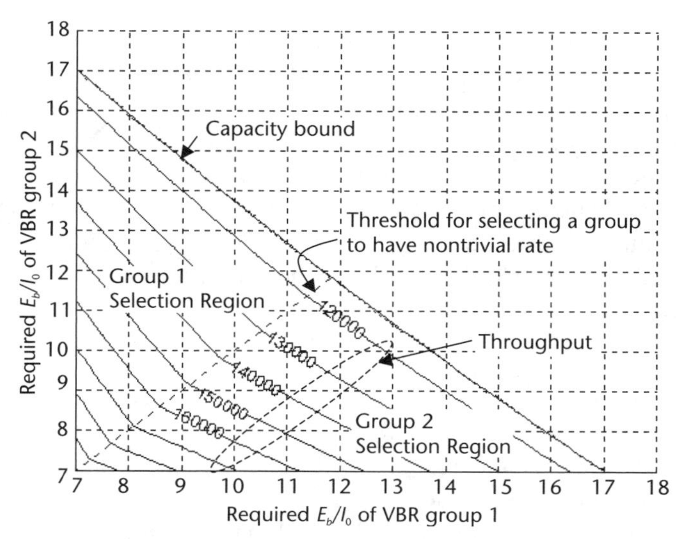

Figure 5.9 Group selection trends and the contour of corresponding throughput according to the difference in the required E_b/I_0 when $R_{v_{1,min}}$ and $R_{v_{2,min}}$ are 9.6 Kbps, and k_{v_1} and k_{v_2} are 6.

density ratio, q_v. In this case, the same values of k_v and $R_{v_{min}}$ are used for both service groups in order to observe group selection trends according to q_v.

The capacity bound in the figure confines the value of E_b/I_0 that can be allowable in the system. From Figure 5.9, it is observed that the group with the smaller value of E_b/I_0 is selected. Further, we know that as the required E_b/I_0 gets smaller, more throughput is achieved.

The trend of group selection of the resource allocation scheme is summarized in Table 5.2. The resource allocation scheme tends to select a group with a smaller number of concurrent users, a larger minimum transmission rate, and smaller required E_b/I_0 when allocating the remaining resources. For smaller k_v, more resources can be allocated to each user by using a certain amount of remaining resources. As $R_{v_{min}}$ becomes larger, users utilize more resources. The selection of a group with smaller k_v and larger $R_{v_{min}}$ is reasonable, as R_v exponentially increases as more resources are allocated, as observed in the previous section. The selection of a group with smaller q_v is also rational, because the increment rate of R becomes larger as q decreases.

5.6 Conclusions

In this chapter, a dynamic resource allocation scheme is proposed to maximize the throughput for multimedia CDMA systems. Because the throughput takes a convex

Table 5.2 Group Selection of the Resource Allocation Scheme for Maximizing the Throughput

Comparison Parameters	*The ith Group Selected*
Number of concurrent users, k_v	$k_{v_i} < k_{v_j}$
Minimum transmission rate, $R_{v_{min}}$	$R_{v_{i,min}} > R_{v_{j,min}}$
Required E_b/I_0, q_v	$q_{v_i} < q_{v_j}$

shape as a function of data rates, more precisely rate factors, the maximum throughput is obtained at one of the cases where the remaining resources are fully allocated to a certain VBR group. This fact reduces the infinite number of possible data rate sets to N possible candidate data rate sets, where N is the number of VBR groups, and makes it feasible to present a simple resource allocation scheme. The proposed allocation scheme provides more average throughput than a scheme allocating the remaining resources to all or several VBR groups and also requires smaller amount of calculation. Thus, this work can be utilized as a method to efficiently utilize the limited system resources.

References

[1] Zander, J., and S. L. Kim, *Radio Resource Management for Wireless Network*, Norwood, MA: Artech House, 2001.

[2] Ortigoza-Guerrero, L., and A. H. Aghvami, *Resource Allocation in Hierarchical Cellular Systems,* Norwood, MA: Artech House, 1999.

[3] Tripathi, N. D., J. H. Reed, and H. F. Van Landingham, *Radio Resource Management in Cellular Systems,* Boston, MA: Kluwer Academic Publishers, 2001.

[4] Sampath, A., P. S. Kumar, and J. M. Holtzman, "Power Control and Resource Management for a Multimedia CDMA Wireless System," *IEEE Proc. of International Symposium on Personal, Indoor and Mobile Radio Communications*, 1995, pp. 21–25.

[5] Yang, Y. R., et al., "Capacity Plane of CDMA System for Multimedia Traffic," *IEEE Electronics Letters*, 1997, pp. 1432–1433.

[6] Ramakrishna, S., and J. M. Holtzman, "A Scheme for Throughput Maximization in a Dual-Class CDMA System," *IEEE Journal on Selected Areas in Communications*, 1998, pp. 830–844.

CHAPTER 6

Voice/Data Mixed CDMA Systems with Prioritized Services

To tackle the RRM issue in CDMA systems supporting multiclass traffic, in this chapter we propose a CAC scheme for CDMA systems supporting voice and data services and analyze the Erlang capacity under the proposed CAC scheme. Service groups are classified by their QoS requirements, such as the required BER and information data rate, and grade of service (GoS) requirements, such as the required call blocking probability. Different traffic types require different system resources based on their QoS requirements. In the proposed CAC scheme, some system resources are reserved exclusively for handoff calls to have higher priority over new calls. Additionally, queuing is allowed for both new and handoff data traffic that are not sensitive to delay. As a performance measure of the suggested CAC scheme, Erlang capacity is introduced. For the performance analysis, a four-dimensional Markov chain model is developed. As a numerical example, Erlang capacity of an IS-95B-type system is depicted, and optimum values of system parameters, such as the number of the reservation channels and queue lengths, are found. Finally, it is observed that Erlang capacity is improved more than two times by properly selecting CAC-related parameters under the proposed CAC scheme. Also, the effect of handoff parameters on the Erlang capacity is observed.

6.1 Introduction

Because future wireless applications will also be more bandwidth intensive and the radio spectrum allocated to wireless communication is hardly able to be extended, the CAC has become an essential network function of wireless networks supporting mixed services. Under a mixed-media CDMA environment, CAC is not a trivial problem.

In [1–3], CAC schemes favoring handoff calls by means of queuing and channel reservation are presented, where some channels are exclusively designated for handoff calls, and a delay-nonsensitive handoff call is put in the queue if the BS finds all channels in the target cell occupied. All of these references focus on voice-oriented FDMA cellular systems. In [4], Pavlidou proposed a mathematical model to analyze the call blocking probability of the mixed voice and data systems when a number of channels is reserved exclusively for handoff calls and only data handoff calls are queued. Furthermore, Calin and Zeghlache suggested a scheme allowing handoff voice calls also to be queued [5]. However, [1–5] are not directly applicable to CDMA systems. Furthermore, it is assumed that voice and data traffic have the same QoS requirements and require same system resources, which is not suitable for the multimedia environments where multimedia traffic requires different system resources based on their QoS requirements.

In this chapter, a CAC scheme for the mixed voice/data CDMA systems supporting the different QoS requirements is proposed. In addition, the Erlang capacity under the proposed CAC scheme is analyzed, where voice and data calls require different system resources based on their QoS requirements, such as the required BER and data transmission rate, respectively. In the proposed CAC scheme, some system resources are reserved exclusively for handoff calls to have higher priority over new calls, and queuing is allowed for both new and handoff data traffic that are not sensitive to delay.

As a performance measure of the proposed CAC scheme, Erlang capacity, defined as a set of the average offered traffic loads of each service group that the CDMA system can carry while the QoS and GoS requirements for all service groups are being satisfied, is utilized so as to consider the performances of all service groups simultaneously. For the performance analysis, we have identified a capacity threshold for voice and data traffic to meet QoS requirements for each kind of traffic and developed a four-dimensional Markov chain model, based on the capacity threshold and the proposed CAC scheme. Furthermore, we have presented the procedure for properly selecting the CAC-related parameters with which the CDMA system can be optimally operated with respect to the system Erlang capacity. As a practical example, an IS-95B-type CDMA system that supports a medium data rate by aggregating multiple codes in the reverse link is considered, and a procedure to select the optimum values of CAC-related parameters, such as the number of the reservation channels and queue size with respect to the Erlang capacity, is illustrated.

The remainder of this chapter is organized as follows: In Section 6.2, we describe the system model. In Section 6.3, a CAC scheme for mixed voice/data CDMA systems is proposed and analyzed, based on the multidimensional Markov model. In Section 6.4, a numerical example is taken into consideration and discussions are given. Finally, conclusions are drawn in Section 6.5.

6.2 System and Traffic Models

6.2.1 System Model

In the case of CDMA, although there is no hard limit on the number of mobile users served, there is a practical limit on the number of simultaneous users in a cell to control the interference between users having the same pilot signal. More specially, in [6–9] the maximum number of current users that CDMA systems can support with QoS requirements was found. As described in (3.1), the system capacity bound of CDMA systems supporting voice and data traffic in the reverse link is expressed as [9]:

$$\gamma_v N_v + \gamma_d N_d \leq 1 \tag{6.1}$$

where

$$\gamma_v = \frac{\alpha}{\dfrac{W}{R_{v_{req}}}\left(\dfrac{E_b}{N_o}\right)_{v_{req}}^{-1}\left\langle\dfrac{1}{1+f}\right\rangle 10^{\frac{Q^{-1}(\beta)}{10}\sigma_x - 0.012\sigma_x^2} + \alpha}$$

$$\gamma_d = \frac{1}{\dfrac{W}{R_{d_{req}}}\left(\dfrac{E_b}{N_o}\right)_{d_{req}}^{-1}\left\langle\dfrac{1}{1+f}\right\rangle 10^{\frac{Q^{-1}(\beta)}{10}\sigma_x - 0.012\sigma_x^2} + 1}$$

All relevant parameters are defined and described in Section 3.1.

The inequality of (6.1) is the necessary and sufficient condition satisfying the system QoS requirements and indicates that calls of different types of services take different amounts of system resources according to their QoS requirements (e.g., information data rate and the required bit energy-to-inference power spectral density ratio). In the following analysis, based on (6.1), we assume that one call attempt of the data service group is equivalent to Λ call attempts of voice service. is defined as $\lfloor \gamma_d / \gamma_v \rfloor$, where $\lfloor x \rfloor$ denotes the greatest integer less than or equal to x. Then, (6.1) can be rewritten as follows.

$$N_v + \Lambda N_d \le \hat{C}_{ETC} \tag{6.2}$$

where $\hat{C}_{ETC} \equiv \lfloor 1 / \gamma_v \rfloor$ is the total number of basic channels within a cell and subscript of "ETC" denotes equivalent telephone (voice) channel (i.e., the voice channel is presumed to the basic channel). Equation (6.2) will be utilized to determine the admission set for the proposed CAC scheme in Section 6.3.

6.2.2 Traffic Model

The considered system employs a circuit switching method to deal with traffic transmission for voice and data calls. Each user shares the system resources with the other users, and it competes with them for use of the system resources. Once a call request is accepted in the system, the call occupies a channel and transmits the information without any delay during call duration. We also assume that two arrivals of voice and data traffic are distributed according to independent Poisson processes with average arrival rate λ_v and λ_d, respectively. In order to consider the fraction of handoff call in a cell, we introduce Λ_h defined as the ratio of handoff traffic to total arrival traffic, and Λ_h is assumed to be controlled as a parameter value. Then, the arrival rates of new voice and handoff voice calls are given by:

$$\lambda_{nv} = (1 - \Lambda_h)\lambda_v, \quad \lambda_{hv} = \Lambda_h \lambda_v \tag{6.3}$$

Similarly, the arrival rates of new data and handoff data calls are given by:

$$\lambda_{nd} = (1 - \Lambda_h)\lambda_d, \quad \lambda_{hd} = \Lambda_h \lambda_d \tag{6.4}$$

Furthermore, we consider rather a simple model to focus on the impact of handoff on the *call level* QoS and system Erlang capacity, and the handoff control mechanism is not considered in detail. That is, for each call, the mobility of the mobile station is modeled by using such parameters as the unencumbered service time and the residence time. The unencumbered service time (the time for which an assigned channel would be held if no handoff is required), T_μ, is assumed to be exponentially distributed with mean $1/\mu$. Here, μ can be μ_v for voice calls or μ_d for data calls. In addition to the unencumbered service time, we also need to define the residence time that a call spends with any BS before handing off to another BS; T_n is the residence time of a new call, and T_b is the residence time of a handoff call. The channel assigned to a call will be held until either the service is completed in the cell of the assignment or the MS moves out of the cell before service completion.

Hence, the channel holding time of a new call, T_{Hn}, and the channel holding time of a handoff call, T_{Hb}, are given as follows:

$$T_{Hn} = \min\left(T_\mu, T_n\right), T_{Hb} = \min\left(T_\mu, T_b\right) \tag{6.5}$$

where "min" indicates the smaller of the two random variables.

Noting (6.5), we can derive the distribution functions of T_{Hn} and T_{Hb} as follows:

$$F_{T_{Hn}}(t) = F_{T_\mu}(t) + F_{T_n}(t)\left[1 - F_{T_\mu}\right] \tag{6.6}$$

$$F_{T_{Hb}}(t) = F_{T_\mu}(t) + F_{T_b}(t)\left[1 - F_{T_\mu}\right] \tag{6.7}$$

Now we assume that T_μ and T_b are exponentially distributed with means $\overline{T}_n = 1/\mu_n$ and $\overline{T}_b = 1/\mu_b$. Then, T_{Hn} and T_{Hb} are also exponentially distributed with $\mu_{Hn} = \mu + \mu_n$ and $\mu_{Hb} = \mu + \mu_b$. Here μ, μ_{Hn}, and μ_{Hb} can be μ_v, μ_{vHn}, and μ_{vHb} for voice calls, or μ_d, μ_{dHn}, and μ_{dHb} for data calls.

Hence, the distribution function of total channel holding time, T_H, in a cell is

$$F_{T_H}(t) = \frac{F_{T_{Hn}}(t)}{1+\gamma_c} + \frac{\gamma_c F_{T_{Hb}}(t)}{1+\gamma_c} \tag{6.8}$$

where γ_c is the ratio of the average handoff attempt rate to the average new arrival attempt rate, and it is given as $\gamma_c = \Lambda_b / (1 - \Lambda_b)$.

Then, the distribution and density functions of T_H are given by:

$$F_{T_H}(t) = F_{T_\mu}(t) + \left(1 - F_{T_\mu}(t)\right)\frac{F_{T_n}(t) + \gamma_c F_{T_b}(t)}{1+\gamma_c} \tag{6.9}$$

$$f_{T_H}(t) = \frac{\mu + \mu_n}{1+\gamma_c} e^{-(\mu_n + \mu)t} + \frac{\gamma_c}{1+\gamma_c}(\mu + \mu_b)e^{-(\mu_b + \mu)t} \tag{6.10}$$

Here, μ and T_H can be μ_v and T_{vH} for voice traffic, or μ_d and T_{dH} for data traffic.

For the following analysis, the distribution of T_H is approximated by an exponential distribution with mean $\overline{T}_H$ [1, 3]. The mean value of T_H, $\overline{T}_H$, is chosen such that the following condition is satisfied:

$$\int_0^{\infty}\left(F_{T_H}^{C} - e^{-\mu_H t}\right)dt = 0 \tag{6.11}$$

where F^{C}_{TH} is the complementary function of F_{TH}.

Then, $\overline{T}_H$ is given as:

$$\overline{T}_H = \frac{1}{\mu_H} = \frac{1}{1+\gamma_c}\left(\frac{1}{\mu+\mu_n} + \frac{\gamma_c}{\mu+\mu_h}\right) \tag{6.12}$$

Especially, the mean values of T_{vH} and T_{dH} for voice and data calls are given as

$$\overline{T}_{vH} = \frac{1}{\mu_{vH}} = \frac{1}{1+\gamma_c}\left(\frac{1}{\mu_v+\mu_n} + \frac{\gamma_c}{\mu_v+\mu_h}\right) \tag{6.13}$$

$$\overline{T}_{dH} = \frac{1}{\mu_{dH}} = \frac{1}{1+\gamma_c}\left(\frac{1}{\mu_d+\mu_n} + \frac{\gamma_c}{\mu_d+\mu_h}\right) \tag{6.14}$$

6.3 Erlang Capacity Analysis Under the Proposed CAC Scheme

In the previous section, we have stipulated a capacity threshold for voice and data traffic in CDMA systems with the concept of the effective bandwidth in order to meet QoS requirements for each kind of traffic, especially in physical layer. In this section, we will propose a CAC scheme based on that capacity threshold. Considering that there are $\hat{C}_{ETC}$ basic channels available in a cell, and one call attempt of data traffic is quantitatively equivalent to the Λ times call attempts of voice traffic in aspects of the system resource, we can design the call admission based on the ideas of reservation and queuing. We propose a CAC scheme as a modification of that in [5], as with Figure 6.1. In Table 6.1, some differences are compared between the proposed scheme and the referred CACs [4, 5]. In particular, [5] considered the buffer for the handoff voice call. However, voice traffic is delay sensitive, and it is not efficient to consider the buffer for the handoff voice call. Subsequently, in the proposed scheme, we consider the buffer for new data call rather than voice call because data traffic is more tolerant of the delay requirement such that some system resources are reserved exclusively for handoff calls to have higher priority over new calls. Queuing is allowed for both new and handoff data traffic that are not sensitive to delay. The full description of the proposed scheme is given as following: Among $\hat{C}_{ETC}$ basic channels, $\hat{C}_{ETC} - C_R$ basic channels are available for new voice, new data, handoff voice, and handoff data calls, while C_R basic channels are reserved exclusively for handoff voice and handoff data calls. In addition, two respective queues with the length of Q_n and Q_h are utilized for new data and handoff data calls, which are not sensitive to time delay, with the principle of *first in first out* (FIFO). That is, if no channel is available in the cell, a new voice call attempt is blocked, and a handoff voice call is forced into termination. On the other hand, new data and handoff data calls go into respective queues with finite length Q_n and Q_h. They will wait until a channel becomes available as long as their associated terminals are in the area covered by the BS of the target cell.

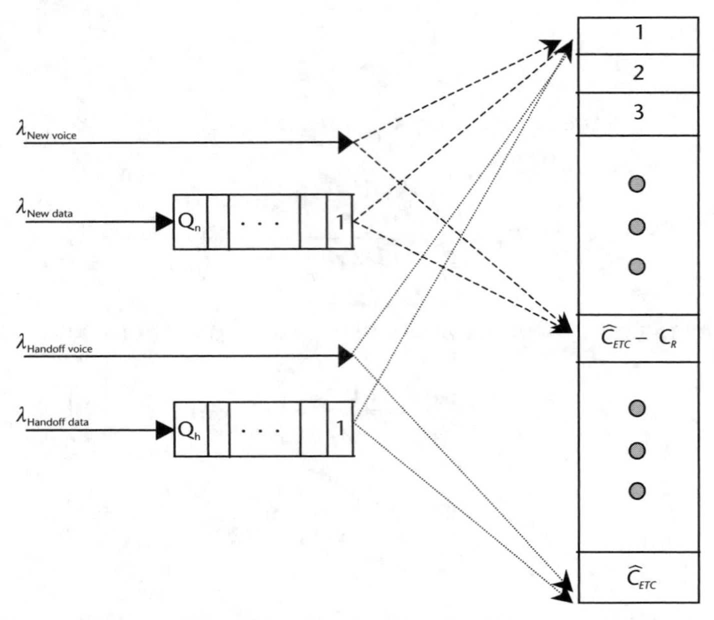

Figure 6.1 Queue system model and channel allocation for the proposed CAC scheme.

Because the waiting time in the queue is restricted only by the mobile residence time in the corresponding cell, the maximum queuing time, T_q, for queued data traffic has the same density function as the mobile residence time in a cell. Hence, T_q has an exponential distribution with $1/\mu_q$. Here, μ_q can be $\mu_{nq}(=\mu_n)$ for the queued new data calls or $\mu_{hq}(=\mu_h)$ or the queued handoff data calls, respectively. Finally, Figure 6.2 summarizes the proposed CAC scheme.

The system performance of the proposed CAC scheme can be analyzed by the birth-death process. For the performance analysis, it is useful to define the occupation state of the cell, S, characterized by the occupation numbers of cells, as a state in the birth-death process such that

Table 6.1 CAC Schemes Based on Reservation and Queuing

CAC Schemes	*Pavlidou's CAC Scheme [4]*	*Calin and Zeghlache's CAC Scheme [5]*	*Proposed CAC Scheme*
New voice call	No reservation and no queuing	No reservation and no queuing	No reservation and no queuing
New data call	No reservation and no queuing	No reservation and no queuing	Calls are queued with finite buffer if the resource is not available
Handoff voice call	Some resources are reserved	Some resources are reserved and calls are queued with finite buffer if the resource is not available	Some resources are reserved
Handoff data call	Some resources are reserved and calls are queued with infinite buffer if the resource is not available	Some resources are reserved and calls are queued with finite buffer if the resource is not available	Some resources are reserved and calls are queued with finite buffer if the resource is not available

```
Once a call is attempted:
IF (sum of used channels after accepting the incoming call ≤ Ĉ_ETC − C_R)
    Incoming call is accepted
ELSE /*not enough basic channels*/
    IF (new call) /*incoming call is new call*/
        IF (new voice call) /*new voice call*/
            Incoming call is blocked
        ELSE /*new data call*/
            IF (number of new data cells in the queue < Q_n)
                Incoming call is inserted in queue
            ELSE
                Incoming call is blocked
    IF /*incoming call is handoff call*/
        IF (sum of used channel after accepting incoming call ≤ Ĉ_ETC)
            Incoming call is accepted
        ELSE /*reservation channel is not enough*/
            IF (handoff voice call) /*handoff voice call*/
                Incoming call is blocked
            ELSE /*handoff data call*/
                IF (number of handoff data calls in the queue < Q_h)
                    Incoming call is inserted in queue
                ELSE
                    Incoming call is blocked
```

Figure 6.2 The proposed CAC algorithm.

$$S = (i, j, m, n)\, i \geq 0,\ j \geq 0,\ i + \Lambda j \leq \hat{C}_{ETC},$$
$$0 \leq m \leq Q_n, \text{and } 0 \leq n \leq Q_n \tag{6.15}$$

where the state variables i and j denote the number of voice and data users in the system, and m and n indicate the number of new and handoff data users in the respective queues.

According to the proposed CAC scheme, a state in the birth-death process falls among the four different admission sets as follows:

$$\begin{aligned}
\Omega_{non-res} &\equiv \left\{(i, j, m, n) | 0 \leq i + \Lambda j \leq \hat{C}_{ETC} - C_R\right\} \\
\Omega_{res} &\equiv \left\{(i, j, m, n) | \hat{C}_{ETC} - C_R < i + \Lambda j \leq \hat{C}_{ETC}\right\} \\
\Omega_{nd-buf} &\equiv \left\{(i, j, m, n) | \hat{C}_{ETC} - C_R - \Lambda < i + \Lambda j \leq \hat{C}_{ETC}, 0 < m \leq Q_n\right\} \\
\Omega_{hd-buf} &\equiv \left\{(i, j, m, n) | \hat{C}_{ETC} - \Lambda < i + \Lambda j \leq \hat{C}_{ETC}, 0 < n \leq Q_h\right\}
\end{aligned} \tag{6.16}$$

The set of all allowable states is given as

$$\Omega_{all} = \Omega_{non-res} \cup \Omega_{res} \cup \Omega_{nd-buf} \cup \Omega_{hd-buf} \tag{6.17}$$

Let $P_{(i,j,m,n)}$ be the probability that four-dimensional Markov chain is in the state $S = (i, j, m, n)$. Then, there is a flow equilibrium balance equation for each state (i.e., the total rate of flowing into a state will be equal to the total rate flowing out from it).

That is,

$$\begin{aligned}
&\text{Rate-In} = \text{Rate-Out} \\
&\text{Rate-In} = \vec{a} \cdot P_{(i+1,j,m,n)} + \vec{b} \cdot P_{(i,j+1,m,n)} + \\
&\vec{c} \cdot P_{(i,j,m+1,n)} + \vec{d} \cdot P_{(i,j,m,n+1)} + \vec{e} \cdot P_{(i-1,j,m,n)} + \\
&\vec{f} \cdot P_{(i,j-1,m,n)} + \vec{g} \cdot P_{(i,j,m-1,n)} + \vec{h} \cdot P_{(i,j,m,n-1)} \\
&\text{Rate-Out} = \left(\vec{i} + \vec{j} + \vec{k} + \vec{l} + \vec{m} + \vec{n} + \vec{o} + \vec{p}\right) \cdot P_{(i,j,m,n)} \\
&\text{for all states}
\end{aligned} \tag{6.18}$$

where the state transitions involved in (6.18) are summarized in the Tables 6.2 and 6.3. The state transition parameters $\vec{a}, \vec{b}, \vec{c}$, and $\vec{d}$ in Table 6.2 occurs when a service is completed, while the parameters $\vec{e}, \vec{f}, \vec{g}$, and $\vec{h}$ occur when a call is admitted in the system. Similarly, the state transition parameters $\vec{i}$, $\vec{j}$, $\vec{k}$, and $\vec{l}$ in Table 6.3 occur when a service is completed, while the parameters $\vec{m}, \vec{n}, \vec{o}$, and $\vec{p}$ occur when a call is admitted in the system.

If the total number of all allowable states is n_s, there are $(n_s - 1)$ linearly independent flow equilibrium balance equations. Based on these $(n_s - 1)$ flow equilibrium balance equations and the normalized equation, $\sum_{(i,j,m,n)\in\Omega_{all}} p_{(i,j,m,n)} = 1$, a set of linear equations of the Markov chain in the form of $\pi Q = P$ can be formed, where π is vector of all states, **Q** is the coefficient matrix of the linear equations, and **P** = [0, ..., 1]. The dimension of π, **Q**, and **P** are $1 \times n_s$, $n_s \times n_s$, $n_s \times 1$, respectively. By solving $\pi = PQ^{-1}$, we obtain all steady-state probabilities.

Based on the proposed CAC scheme, the call attempts of new data and handoff voice calls are blocked if there is no channel available. Hence, the call blocking probabilities for new voice and handoff voice calls are given as follows:

$$P_{(B,nv)} = \sum_{s\in\Omega_{(B,nv)}} P_{(i,j,m,n)} \tag{6.19}$$

$$P_{(B,hv)} = \sum_{s\in\Omega_{(B,hv)}} P_{(i,j,m,n)} \tag{6.20}$$

where

$$\begin{aligned}
\Omega_{(B,nv)} &= \left\{(i,j,m,n) | \hat{C}_{ETC} - C_R < i + \Lambda j \le \hat{C}_{ETC}\right\} \\
\Omega_{(B,hv)} &= \left\{(i,j,m,n) | i + \Lambda j = \hat{C}_{ETC}\right\}
\end{aligned}$$

Table 6.2 The State Transition Rates Related with Rate-In Flow

Parameter	*Definition*	*Value*
$\vec{a}$	$(i+1,j,m,n)\xrightarrow{\vec{a}}(i,j,m,n)$	$\vec{a}=\begin{cases}(i+1)\mu_{vH} & \text{if } (i+1,j,m,n)\in\Omega_{non-res}\\ \hat{i}\mu_{vH}+\left(i+1-\hat{i}\right)\mu_{vHh} & \text{if } (i+1,j,m,n)\in\Omega_{res}\\ 0 & \text{otherwise}\end{cases}$ where $\hat{i}=\hat{C}_{etc}-C_R-\Lambda\cdot j$
$\vec{b}$	$(i,j+1,j,m,n)\xrightarrow{\vec{b}}(i,j,m,n)$	$\vec{b}=\begin{cases}(i+1)\mu_{dH} & \text{if } (i,j+1,m,n)\in\Omega_{non-res}\\ \hat{j}\mu_{dH}+\left(j+1-\hat{j}\right)\mu_{dHh} & \text{if } (i,j+1,m,n)\in\Omega_{res}\\ 0 & \text{otherwise}\end{cases}$ where $\hat{j}=\left\lfloor\left(\hat{C}_{ETC}-C_R-i\right)/\Lambda\right\rfloor$
$\vec{c}$	$(i,j,m+1,n)\xrightarrow{\vec{c}}(i,j,m,n)$	$\vec{c}=\begin{cases}\hat{j}\mu_{dH}+\left(j-\hat{j}\right)\mu_{dHh}+(m+1)\mu_{qn} & \text{if } (i,j,m+1,n)\in\Omega_{nd-buf}\\ 0 & \text{otherwise}\end{cases}$ where $\hat{j}=\left\lfloor\left(\hat{C}_{ETC}-C_R-i\right)/\Lambda\right\rfloor$
$\vec{d}$	$(i,j,m,n+1)\xrightarrow{\vec{d}}(i,j,m,n)$	$\vec{d}=\begin{cases}\hat{j}\mu_{dH}+\left(j-\hat{j}\right)\mu_{dHh}+(n+1)\mu_{qh} & \text{if } (i,j,m,n+1)\in\Omega_{hd-buf}\\ 0 & \text{otherwise}\end{cases}$ where $\hat{j}=\left\lfloor\left(\hat{C}_{ETC}-C_R-i\right)/\Lambda\right\rfloor$
$\vec{e}$	$(i-1,j,m,n)\xrightarrow{\vec{e}}(i,j,m,n)$	$\vec{e}=\begin{cases}\lambda_{nv}+\lambda_{hv} & \text{if } (i,j,m,n)\in\Omega_{non-res}\\ \lambda_{hv} & \text{if } (i,j,m,n)\in\Omega_{res}\\ 0 & \text{otherwise}\end{cases}$
$\vec{f}$	$(i,j-1,m,n)\xrightarrow{\vec{f}}(i,j,m,n)$	$\vec{f}=\begin{cases}\lambda_{nd}+\lambda_{hd} & \text{if } (i,j,m,n)\in\Omega_{non-res}\\ \lambda_{hd} & \text{if } (i,j,m,n)\in\Omega_{res}\\ 0 & \text{otherwise}\end{cases}$
$\vec{g}$	$(i,j,m-1,n)\xrightarrow{\vec{g}}(i,j,m,n)$	$\vec{g}=\begin{cases}\lambda_{nd} & \text{if } (i,j,m,n)\in\Omega_{nd-buf}\\ 0 & \text{otherwise}\end{cases}$
$\vec{h}$	$(i,j,m,n-1)\xrightarrow{\vec{h}}(i,j,m,n)$	$\vec{h}=\begin{cases}\lambda_{hd} & \text{if } (i,j,m,n)\in\Omega_{hd-buf}\\ 0 & \text{otherwise}\end{cases}$

On the other hand, new and handoff data calls are blocked if there is no channel available, and the respective queue is also full. That is, if all channels are busy, but there is at least one place unoccupied in the queue, then new and handoff calls are inserted into the respective queues to wait for service. However, if the waiting time exceeds the maximum queuing time before they get a channel, they will be blocked.

Let $P_{(full,nd)}$ and $P_{(full,hd)}$ denote the probability that new and handoff data calls find the respective queues are full, respectively. Then, $P_{(full,nd)}$ and $P_{(full,hd)}$ are given as follows:

$$P_{(full,nd)}=\sum_{s\in\Omega_{(full,nd)}}P_{(i,j,m,n)} \tag{6.21}$$

$$P_{(full,hd)}=\sum_{s\in\Omega_{(full,hd)}}P_{(i,j,m,n)} \tag{6.22}$$

where

Table 6.3 The State Transition Rates Related with Rate-Out Flow

Parameter	*Definition*	*Value*
$\vec{i}$	$(i,j,m,n)\xrightarrow{\vec{i}}(i-1,j,m,n)$	$\vec{i}=\begin{cases} i\mu_{vH} & \text{if } (i,j,m,n)\in\Omega_{non-res} \\ \hat{i}\mu_{vH}+(i-\hat{i})\mu_{vHb} & \text{if } (i,j,m,n)\in\Omega_{res} \\ 0 & \text{otherwise} \end{cases}$ where $\hat{i}=\hat{C}_{ETC}-C_R-K\cdot j$
$\vec{j}$	$(i,j,m,n)\xrightarrow{\vec{j}}(i,j-1,m,n)$	$\vec{j}=\begin{cases} j\mu_{dH} & \text{if } (i,j,m,n)\in\Omega_{non-res} \\ \hat{j}\mu_{dH}+(j-\hat{j})\mu_{dHb} & \text{if } (i,j,m,n)\in\Omega_{res} \\ 0 & \text{otherwise} \end{cases}$ where $\hat{j}=\left\lfloor(\hat{C}_{ETC}-C_R-i)/K\right\rfloor$
$\vec{k}$	$(i,j,m,n)\xrightarrow{\vec{k}}(i,j,m-1,n)$	$\vec{k}=\begin{cases} \hat{j}\mu_{dH}+(j-\hat{j})\mu_{dHb}+m\mu_{qn} & \text{if } (i,j,m,n)\in\Omega_{hd-buf} \\ 0 & \text{otherwise} \end{cases}$
$\vec{l}$	$(i,j,m,n)\xrightarrow{\vec{l}}(i,j,m,n-1)$	$\vec{l}=\begin{cases} \hat{j}\mu_{dH}+(j-\hat{j})\mu_{dHb}+n\mu_{qn} & \text{if } (i,j,m,n-1)\in\Omega_{hd-buf} \\ 0 & \text{otherwise} \end{cases}$
$\vec{m}$	$(i,j,m,n)\xrightarrow{\vec{m}}(i+1,j,m,n)$	$\vec{m}=\begin{cases} \lambda_{nv}+\lambda_{hv} & \text{if } (i+1,j,m,n)\in\Omega_{non-res} \\ \lambda_{hv} & (i+1,j,m,n)\in\Omega_{res} \\ 0 & \text{otherwise} \end{cases}$
$\vec{n}$	$(i,j,m,n)\xrightarrow{\vec{n}}(i,j+1,m,n)$	$\vec{n}=\begin{cases} \lambda_{nd}+\lambda_{hd} & \text{if } (i,j+1,m,n)\in\Omega_{non-res} \\ \lambda_{hd} & \text{if } (i,j+1,m,n)\in\Omega_{res} \\ 0 & \text{otherwise} \end{cases}$
$\vec{o}$	$(i,j,m,n)\xrightarrow{\vec{o}}(i,j,m+1,n)$	$\vec{o}=\begin{cases} \lambda_{nd} & \text{if } (i,j,m+1,n)\in\Omega_{nd-buf} \\ 0 & \text{otherwise} \end{cases}$
$\vec{p}$	$(i,j,m,n)\xrightarrow{\vec{p}}(i,j,m,n+1)$	$\vec{p}=\begin{cases} \lambda_{hd} & \text{if } (i,j,m,n+1)\in\Omega_{hd-buf} \\ 0 & \text{otherwise} \end{cases}$

$$\Omega_{(full,nd)}=\left\{(i,j,m,n)|\,\hat{C}_{ETC}-C_R\Lambda<i+\Lambda j\le\hat{C}_{ETC},m=Q_n,0\le n\le Q_h\right\}$$

$$\Omega_{(full,hd)}=\left\{(i,j,m,n)|\,\hat{C}_{ETC}-\Lambda<i+\Lambda j\le\hat{C}_{ETC},0\le m\le Q_n,n=Q_h\right\}$$

Also, the handoff failure probability for the new and handoff data calls due to their time outs are provided respectively by the following equations.

$$P_{(F,nd)}=\frac{\sum_{s\in\Omega_{(nd-buf)}}k\mu_{qn}P_{(i,j,k,l)}}{\lambda_{nd}\left(1-P_{(full,nd)}\right)} \tag{6.23}$$

$$P_{(F,hd)}=\frac{\sum_{s\in\Omega_{(hd-buf)}}l\mu_{qh}P_{(i,j,k,l)}}{\lambda_{hd}\left(1-P_{(full,hd)}\right)} \tag{6.24}$$

Finally, total call blocking probabilities for all new and handoff data traffic are given as:

$$P_{(B,nd)} = P_{(F,nd)}\left(1 - P_{(full,nd)}\right) + P_{(full,nd)} \tag{6.25}$$

$$P_{(B,hd)} = P_{(F,hd)}\left(1 - P_{(full,hd)}\right) + P_{(full,hd)} \tag{6.26}$$

In this chapter, as a performance measure for the proposed CAC scheme, Erlang capacity is introduced. It is defined as a set of average loads of voice and data traffic that can be supported with a given quality and availability of service. In this case, Erlang capacity is given as:

$$C_{Erlang} \equiv \{(\hat{\rho}_v, \hat{\rho}_d)\} = \left\{ \begin{aligned} &(\rho_v, \rho_d) | P_{(B,nv)} \leq P_{(B,nv)_{req}}, P_{(B,hv)} \leq P_{(B,hv)_{req}}, \\ &P_{(B,nd)} \leq P_{(B,nd)_{req}}, and\ P_{(B,hd)} \leq P_{(B,hd)_{req}} \end{aligned} \right\}$$

where $\rho_v = \lambda_v/\mu_{vH}, \rho_d = \lambda_d/\mu_{dH}, \lambda_v$ and λ_d are the call arrival rates of voice and date calls per cell, respectively; $1/\mu_{vH}$ and $1/\mu_{dH}$ are the average total channel holding times of voice and data calls, respectively; and $P_{(B,nv)_{req}}$, $P_{(B,nd)_{req}}$, $P_{(B,hv)_{req}}$, and $P_{(B,hd)_{req}}$ are the required call blocking probabilities of new voice, new data, handoff voice, and handoff data calls, respectively.

The system Erlang capacity is the set of values of $\{(\hat{\rho}_v, \hat{\rho}_d)\}$ that keeps the call-blocking probability experimented by each traffic less than the required call blocking probability of each traffic call, which is typically given as 1% for new calls and 0.1% for handoff calls. In this situation, the Erlang capacity, with respect to each call, can be calculated as a function of offered loads of voice and data traffic, by contouring the call blocking probability experimented by each traffic at the level of the required call blocking probability. Furthermore, total system Erlang capacity is determined by the overlapped region of Erlang capacities with respect to each call. An easy way to visualize total system Erlang capacity is to consider the overlapped Erlang capacity region as total system Erlang capacity.

A general goal of the proposed CAC scheme is to carry the largest Erlang capacity for a given amount of spectrum and further to find the optimum values of system parameters, such as the number of the reservation channels and queue size with respect to the Erlang capacity.

6.4 Numerical Example

As a numerical example, let's consider a typical IS-95B CDMA system supporting voice and data traffic. IS-95B systems support medium data rates by aggregating multiple codes in both directions, to and from the mobile devices, without changing the IS-95 air interface, and maintaining compatibility with existing BS hardware [10]. The system parameters under the consideration are shown in Table 6.4. In the case of numerical example, Λ and $\hat{C}_{ETC}$ are given as 4 and 27, based on (6.2). It means that there are 29 basic channels, and one call attempt of data traffic is quantitatively equivalent to four call attempts of voice traffic. Also, we assume that all MSs stay in a cell for 1,800 seconds; the average unencumbered service time is 200 seconds for

Table 6.4 System Parameters for the Numerical Example

Parameters	*Symbol*	*Value*
Allocated frequency bandwidth	W	1.25 Mbps
Required bit transmission rate for voice traffic	R_v	9.6 Kbps
Required bit transmission rate for data traffic	R_d	19.2 Kbps
Required bit energy-to-interference power spectral density ratio for voice traffic	$\left(\frac{E_b}{N_o}\right)_{v\ req}$	7 dB
Required bit energy-to-interference power spectral density ratio for data traffic	$\left(\frac{E_b}{N_o}\right)_{d\ req}$	7 dB
System reliability requirement	$\beta\%$	99%
Frequency reuse factor	$\left\langle\frac{1}{1+f}\right\rangle$	0.7
Standard deviation of received SIR	σ_x	1 dB
Voice activity factor	α	3/8

both services; the maximum queuing times of new and handoff data calls are 1,800 seconds, respectively; and Λ_h is 0.2. The average call arrival rates of voice and data, λ_v and λ_d, are variable. Because Λ_h is given as 0.2, the average arrival rates of new voice, handoff voice, new data, and handoff data calls are $0.8\lambda_v$, $0.2\lambda_v$, $0.8\lambda_v$, and $0.2\lambda_v$, respectively. The traffic-related parameters are summarized in Table 6.5.

Figure 6.3 shows the Erlang capacity region that the system can support with 1% call blocking probability for new calls and 0.1% for handoff calls when $C_R = 0$, $Q_n = 0$, and $Q_h = 0$. This case is conceptually correspondent to the complete sharing scheme without considering any priority of calls. It means that a call request is blocked if and only if there are not sufficient resources to service that call. From Figure 6.3, we observe two facts. The first is that data traffic has more impact than voice traffic on Erlang capacity because the effective bandwidth required by one data user is larger than that of one voice user. That is, the Erlang capacity regions limited by the required call blocking probabilities of new and handoff data calls are smaller than those limited by the required call blocking probabilities of new and

Table 6.5 Traffic Parameters for the Numerical Example

Parameters	*Symbol*	*Value*
Average unencumbered service time for voice call	$1/\mu_v$	200 seconds
Average unencumbered service time for data call	$1/\mu_d$	200 seconds
Average residence time for new call	$1/\mu_n$	1,800 seconds
Average residence time for handoff call	$1/\mu_h$	1,800 seconds
Maximum queuing time for new data call	$1/\mu_{qn}$	1,800 seconds
Maximum queuing time for handoff data call	$1/\mu_{qh}$	1,800 seconds
Ratio of handoff traffic to total arrival traffic	Λ_h	0.2
Average arrival time for data call	$1/\lambda_d$	Variable
Average arrival time for voice call	$1/\lambda_v$	Variable
Required call blocking probabilities for new voice and new data calls	$P_{(B,nv)_{req}}$ $P_{(B,nd)_{req}}$	1%
Required call blocking probabilities for handoff voice and handoff data calls	$P_{(B,hv)_{req}}$ $P_{(B,hd)_{req}}$	0.1%

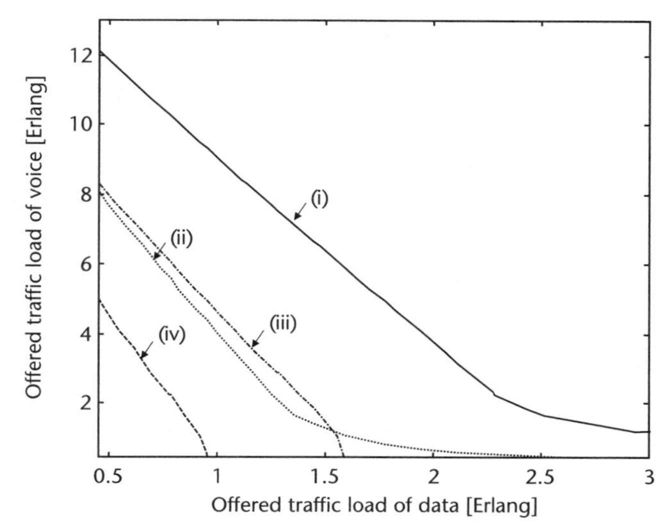

Figure 6.3 Erlang capacity when $C_R = 0$, $Q_n = 0$, and $Q_h = 0$. The curve represented by (i) is the Erlang capacity limited by the required call blocking probability of new voice calls (1%); the curve represented by (ii) is the Erlang capacity limited by the required call blocking probability of handoff voice calls (0.1%); the curve represented by (iii) is the Erlang capacity limited by the required call blocking probability of new data calls (1%); and the curve represented by (iv) is the Erlang capacity limited by the required call blocking probability of handoff data calls (0.1%).

handoff voice calls. The other fact is that total system Erlang capacity region is mainly determined by the Erlang capacity limited by the required call blocking probability of handoff data calls, as the system should satisfy the required call blocking probabilities of all service groups simultaneously.

Hence, it is required to get a proper tradeoff between Erlang capacities that are limited by the required call blocking probabilities of all traffic groups so as to enhance total system Erlang capacity. This observation leads us to the operation of the proposed CAC scheme.

Figure 6.4 shows the effect of the number of the reservation channels, C_R, on Erlang capacity. In this case, some channels are exclusively reserved for voice and data handoff calls, which is very useful, especially when both voice and data traffic are in real time and sensitive to delay. The main observation point is to find the optimal number of the reservation channels with respect to the Erlang capacity. As we see in Figure 6.4, Erlang capacity regions that are limited by the required call blocking probabilities of handoff voice and data calls increase, respectively, as the number of the reservation channels for handoff calls increases—see (ii) and (iv) in Figure 6.4. On the other hand, Erlang capacity regions that are limited by the required call blocking probabilities of new voice and new data calls decrease respectively—see (i) and (iii) in Figure 6.4. In particular, total system Erlang capacity is determined by Erlang capacity limited by the required blocking probability of handoff data calls until three basic channels are reserved for handoff calls.

When more than three basic channels are reserved for handoff calls, then total system Erlang capacity will be determined by the Erlang capacity limited by the required call blocking probability of new data calls. However, we can observe that total system Erlang capacity increases when reserving four basic channels for handoff calls by comparing Figures 6.3 and 6.4(d). Also, Figure 6.4 shows that it is

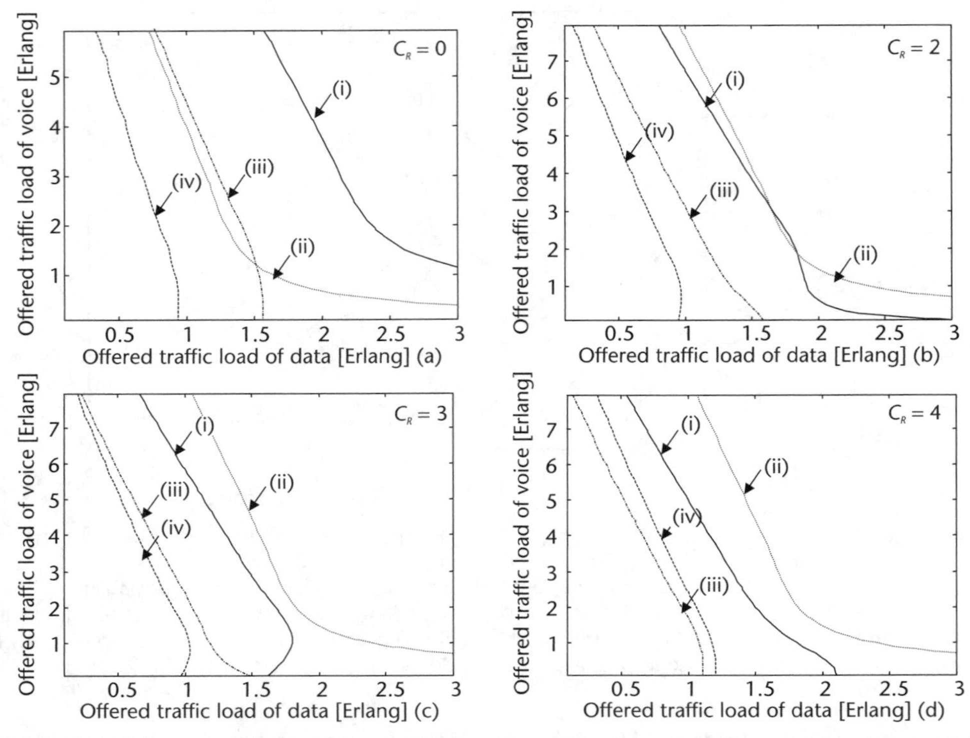

Figure 6.4 Erlang capacity according to the number of the reservation channels for voice and data handoff calls when $Q_n = 0$ and $Q_h = 0$: (a) $C_R = 1$, (b) $C_R = 2$, (c) $C_R = 3$, and (d) $C_R = 4$. For each case, the curve represented by (i) is the Erlang capacity limited by the required call blocking probability of new voice calls (1%); the curve represented by (ii) is the Erlang capacity limited by the required call blocking probability of handoff voice calls (0.1%); the curve represented by (iii) is the Erlang capacity limited by the required call blocking probability of new data calls (1%); and the curve represented by (iv) is the Erlang capacity limited by the required call blocking probability of handoff data calls (0.1%).

inefficient to reserve more than four basic channels for handoff calls by which Erlang capacity limited by the required call blocking probability of new data calls will be more restricted. Hence, in the case where only a reservation scheme is considered, the optimum value of the number of the reservation channels for handoff calls is four.

In the proposed CAC scheme, two respective queues with the finite queue length of Q_n and Q_h are utilized for new and handoff data calls, respectively. Figure 6.5 shows the effect of the length of respective queues for new and handoff data calls on Erlang capacity. As we see in Figure 6.5, Erlang capacity regions that are limited by the required call blocking probabilities of new and handoff data calls increase as the length of queues for new and handoff calls get larger—see (iii) and (iv) in Figure 6.5. On the other hand, Erlang capacity regions that are limited by the required call blocking probabilities of new and handoff voice calls are not affected by the respective queues—see (i) and (ii) in Figure 6.5. Here, we consider the case where the number of the reservation channels for handoff traffic is two. The reason is that the Erlang capacity region that is overlapped by Erlang capacities limited by the required call blocking probabilities of new and handoff voice calls is maximized when $C_R = 2$—see (i) and (ii) in Figure 6.4(b). In addition, Erlang capacities that are limited by the required call blocking probability of new and handoff data calls can

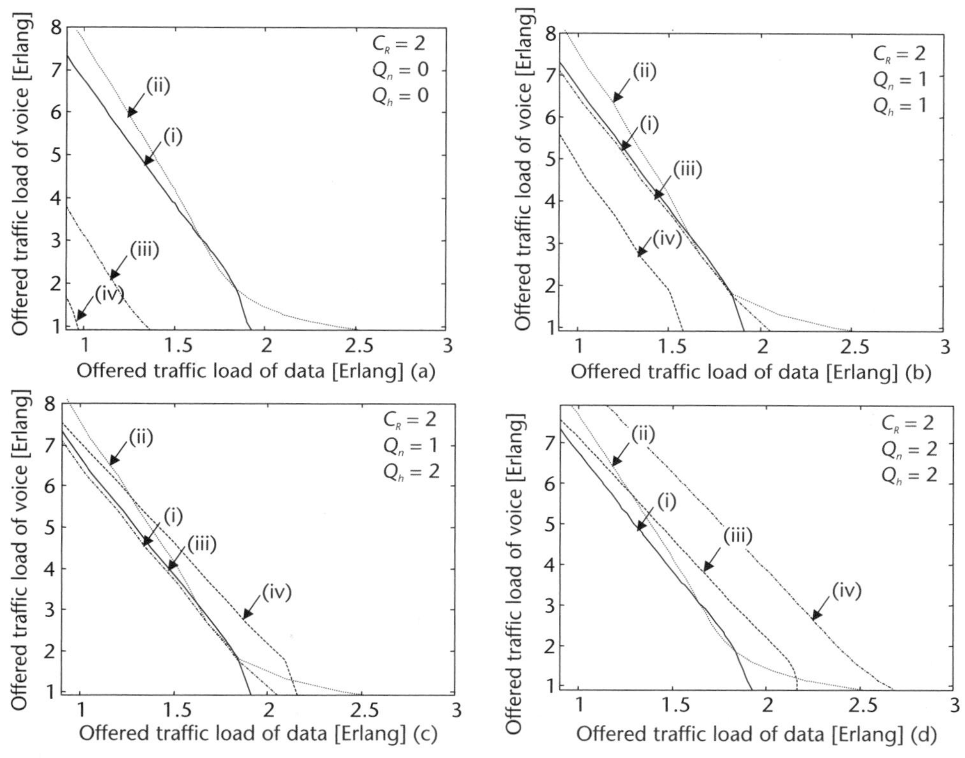

Figure 6.5 Erlang capacity according to the length of the queue for new and handoff data calls when $C_R = 2$: (a) $Q_n = 0$ and $Q_h = 0$ (b) $Q_n = 1$ and $Q_h = 1$, (c) $Q_n = 1$ and $Q_h = 2$, and (d) $Q_n = 2$ and $Q_h = 2$. For each case, the curve represented by (i) is the Erlang capacity limited by the required call blocking probability of new voice calls (1%); the curve represented by (ii) is the Erlang capacity limited by the required call blocking probability of handoff voice calls (0.1%); the curve represented by (iii) is the Erlang capacity limited by the required call blocking probability of new data calls (1%); and the curve represented by (iv) is the Erlang capacity limited by the required call blocking probability of handoff data calls (0.1%).

be adjusted through the queue length. Finally, Figure 6.5 shows that total system Erlang capacity is maximized when $C_R = 2$, $Q_n = 2$, and $Q_h = 2$. Furthermore, it is observed that total Erlang capacity under the proposed CAC is increased more than two times, comparing Figure 6.5(d) with Figure 6.3. It is noteworthy that total system Erlang capacity is not increased even if the length of respective queues is increased more than two times because the Erlang capacity limited by the required call blocking probability of new voice calls is a dominant factor, which determines total system Erlang capacity. Also, the queuing time delay is introduced due to the queue. The larger the queue length gets, the longer the time delay of the queue. Hence, the optimum values of the length of respective queues for new and handoff calls and the number of reservation channels are two, two, and two respectively, with respect to both Erlang capacity and queuing time delay.

Figure 6.6 shows the Erlang capacity according to the changes of Λ_h when $C_R = 2$, $Q_n = 2$, and $Q_h = 2$. As Λ_h gets higher, the handoff arrival rates of voice and data will be higher. Figure 6.6 shows that the Erlang capacities limited by the required call blocking probabilities of handoff voice and handoff data calls decrease as Λ_h increases—see (ii) and (iv) in Figure 6.6. On the other hand, the Erlang capacities

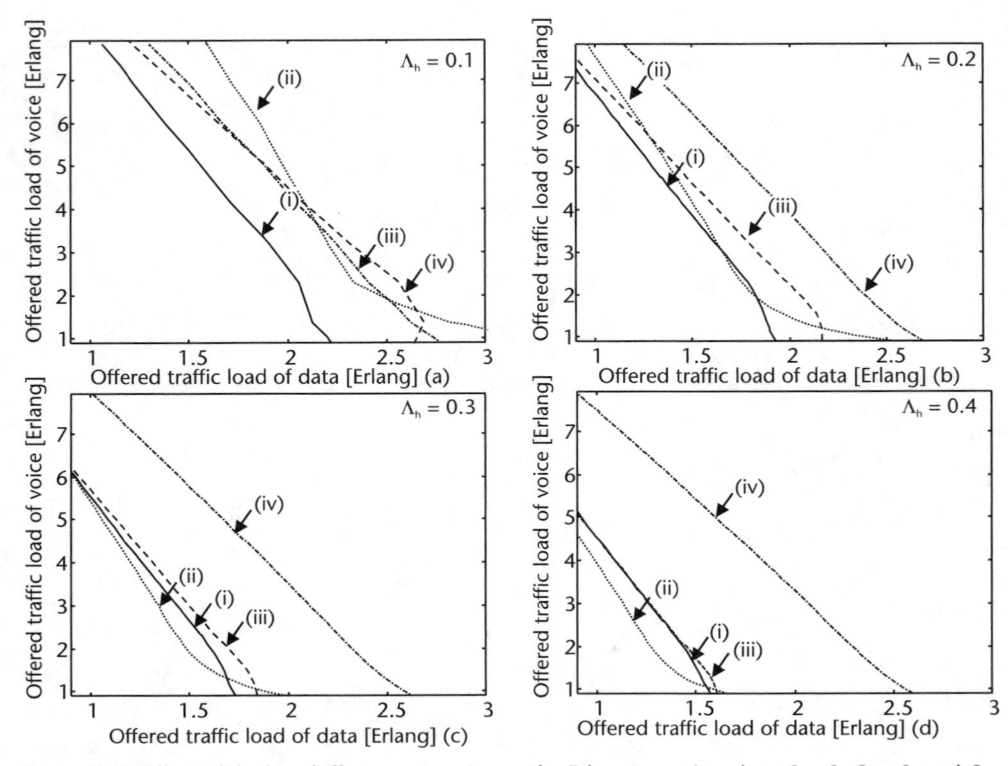

Figure 6.6 Effect of the handoff parameter, Λ_h, on the Erlang capacity when $C_R = 2$, $Q_n = 2$, and $Q_h = 2$: (a) $\Lambda_h = 0.1$, (b) $\Lambda_h = 0.2$, (c) $\Lambda_h = 0.3$, and (d) $\Lambda_h = 0.4$. For each case, the curve represented by (i) is the Erlang capacity limited by the required call blocking probability of new voice calls (1%); the curve represented by (ii) is the Erlang capacity limited by the required call blocking probability of handoff voice calls (0.1%); the curve represented by (iii) is the Erlang capacity limited by the required call blocking probability of new data calls (1%); and the curve represented by (iv) is the Erlang capacity limited by the required call blocking probability of handoff data calls (0.1%).

limited by the required call blocking probabilities of new voice and new data calls increase as Λ_h increases—see (i) and (iii) in the figure. Finally, Figure 6.6 shows total system Erlang capacity decreases with the increase of Λ_h.

The optimum values of C_R, Q_n, and Q_h should be readjusted to increase total Erlang capacity. When $\Lambda_h = 0.1$, for example, reservation channels less than two basic channels are enough to obtain the maximized total Erlang capacity. When $\Lambda_h = 0.3$ or 0.4, more reservation channels are necessary for handoff calls. Finally, it is noteworthy that although only the effect of Λ_h on the Erlang capacity has been considered, the effect of the other handoff parameters such as the residence time, the maximum queuing time on the Erlang capacity, can be observed through a way similar to the case of Λ_h.

6.5 Conclusion

In this chapter, we have proposed and analyzed a CAC scheme for a mixed voice/data CDMA system in order to accommodate more system Erlang capacity. In the proposed scheme, some system resources are reserved exclusively for handoff calls to have higher priority over new calls. Additionally, the queuing is allowed for both new and handoff data traffic that are not sensitive to delay. For the performance analysis, a four-dimensional Markov chain model is developed. Through a numerical example of the Erlang capacity for an IS-95B-type system, we observe that data users have more impact on the Erlang capacity than voice users because the effective bandwidth of one data user is larger than that of one voice user.

It is also observed that the Erlang capacities with respect to all traffic groups should be balanced to enhance total system Erlang capacity. Subsequently, there are optimal values of reservation channels and queue lengths in order to maximize total Erlang capacity. In the case where only a reservation scheme is considered, the optimum value of the number of the reservation channels for handoff calls is four with respect to the Erlang capacity. On the other hand, for the case in which the queue and reservation schemes are combined, the optimum values of the number of the reservation channels for handoff calls and the length of respective queues for new and handoff data calls are two, two, and two, respectively, where the Erlang capacity is improved more than two times.

References

[1] Hong, D., and S. Rappaport, "Traffic Model and Performance Analysis for Cellular Mobile Radio Telephone Systems with Prioritized and Nonprioritized Handoff Procedures," *IEEE Trans. on Vehicular Technology*, 1986, pp. 77–92.

[2] Del Re, E., et al., "Handover and Dynamic Channel Allocation Techniques in Mobile Cellular Networks," *IEEE Trans. on Vehicular Technology*, 1995, pp. 229–237.

[3] Hong, D., and S. Rappaport, "Priority Oriented Channel Access for Cellular Systems Serving Vehicular and Portable Radio Telephones," *IEE Proc. of Commun.*, 1989, pp. 339–346.

[4] Pavlidou, F., "Two-Dimensional Traffic Models for Cellular Mobile Systems," *IEEE Trans. on Commun.*, 1994, pp. 1505–1511.

[5] Calin, D., and D. Zeghlache, "Performance and Handoff Analysis of an Integrated Voice-Data Cellular System," *IEEE Proc. of PIMRC*, 1997, pp. 386–390.

[6] Sampath, A., P. S. Kumar, and J. M. Holtzman, "Power Control and Resource Mangement for a Multimedia CDMA Wireless System," *IEEE Proc. of International Symposium on Personal, Indoor, and Mobile Radio Communications,* 1995, pp. 21–25.

[7] Gilhousen, K. S., et al., "On the Capacity of a Cellular CDMA System," *IEEE Trans. on Vehicular Technology,* 1991, pp. 303–312.

[8] Yang, Y. R., et al., "Capacity Plane of CDMA System for Multimedia Traffic," *IEE Electrononics Letters*, 1997, pp. 1432–1433.

[9] Koo, I., et al., "A Generalized Capacity Formula for the Multimedia DS-CDMA System," *IEEE Proc. of Asia-Pacific Conference on Communication,* 1997, pp. 46–50.

[10] IS-95-B, "Mobile Station-Base Station Compatibility Standard for Dual-Mode Wideband Spread Spectrum Cellular System," 1999.

CHAPTER 7

Erlang Capacity of CDMA Systems Supporting Multiclass Services

In FDMA and TDMA systems, traffic channels are allocated to calls as long as they are available. Incoming calls are blocked when all channels have been assigned.

The physical parallel in CDMA systems is for a call to arrive and find that the BS has no receiver processors left to serve it [1]. In a CDMA system, the CE in each BS corresponds to the receiver processor and performs the baseband spread spectrum signal processing of a received signal for a given channel (pilot, sync, paging, or traffic channel). Practically, CDMA systems are equipped with a finite number of CEs with a cost-efficient strategy because CEs are a cost part of BSs, which introduce inherent hard blocking in CDMA systems. However, often a more stringent limit on the number of simultaneous users in a CDMA system is the total interference created by the admitted users, and its measurement is the outage, which occurs when the interference level reaches a predetermined value above the background noise level. In this situation, a call attempt in CDMA systems can be blocked not only by the maximum number of supportable users in the air link but also by the maximum number of CEs available in BS, and the Erlang capacity will be confined by these two resource limits.

In this book, we tackle the Erlang capacity evaluation of CDMA systems with following two cases: the first one is that there is a finite number of CEs in a BS, and the second one is that there is infinite number of CEs in a BS.

First, this chapter will deal with the Erlang capacity of CDMA systems supporting multiclass services when there is no limitation of the CEs in a BS, and Chapter 8 will also be devoted to the capacity evaluation of CDMA system supporting voice and data services under the delay constraint. After that, the remaining chapters will be devoted to the capacity evaluation of CDMA systems with consideration of both the limitation on the maximum number of CEs available in a BS and the limitation on the maximum number of supportable simultaneous users in an air link.

7.1 Introduction

Over the past decade, wireless communication networks have experienced tremendous development. Future wireless networks will expand their services from voice to mobile systems and from data services to multimedia services, such as voice, data, graphics, and low-resolution video using advanced multiple access techniques [2–4].

Many studies have been devoted to supporting multimedia services in CDMA systems. In particular, the research to find the maximum current number of users (defined as the system capacity here) that CDMA systems can support in the reverse link has been done in [3, 5–7]. For the purpose of controlling the system, more than estimating supportable size of the system at an instant, another measure of the system capacity is peak load that can be supported with a given quality and with availability of service as measured by the blocking probability. The average traffic load in terms of the average number of users requesting service resulting in this blocking probability is called as the Erlang capacity. In [8], Viterbi and Viterbi reported the Erlang capacity of CDMA systems for only voice calls, based on outage probability. The outage probability is defined as the probability that the interference plus noise power density I_o exceeds the noise power density N_o by a factor $1/\eta$, where η takes on typical values between 0.25 and 0.1 [8]. Also, Viterbi and Viterbi presumed outage probability to call blocking probability. Call blocking is however mainly caused when a call is controlled by a CAC rule, and the outage probability is not directly correspondent to the call blocking.

In contrast with [8], we will in this chapter extend the analysis of Erlang capacity to case of CDMA systems supporting multiclass services, based on a multidimension *M/M/m* loss model. For the reference of CAC, a system capacity bound with respect to the maximum number of simultaneous users is utilized. With the model, the call blocking probability is given by the well-known Erlang B formula. Furthermore, the channel reservation concept is adopted to increase total system Erlang capacity by making the Erlang capacities with respect to voice and data calls be balanced.

The remainder of this chapter is organized as follows. In Section 7.2, we briefly summarize the system capacity of a multimedia CDMA system from the viewpoint of maximum concurrent number of users. We then stipulate it as a CAC rule. In Section 7.3, we present an analytical approach for evaluating the call blocking probability and Erlang capacity. In Section 7.4, a numerical example is taken into consideration and the channel reservation scheme is also considered to increase total system Erlang capacity. Finally, in Section 7.5, some conclusions are drawn.

7.2 System Model and System Capacity

Regarding the evaluation of Erlang capacity, Viterbi and Viterbi reported the Erlang capacity of CDMA system for voice calls only. This was based on outage probability, where the outage probability is defined as the probability that the interference plus noise power density I_o exceeds the noise power density N_o by a factor $1/\eta$, where η takes on typical values between 0.25 and 0.1 [8]. Viterbi's model for Erlang capacity is a $M/M/\infty$ queue with voice activity factor $\rho(\rho \approx 0.4)$ (i.e., a queue model with Poisson input and with infinite service channels having IID exponential service time distribution is considered, where M and M means that each user has exponentially distributed interarrival times and service times and ∞ means an infinite number of available servers). More fundamental explanations on $M/M/\infty$ queue are available in Appendix A. Because the capacity of a CDMA system is soft, Viterbi and Viterbi prefer outage probability to blocking probability. The resulting expression for outage probability is simply the tail of the Poisson distribution [8, 9]

$$P_{out} < e^{-\frac{\rho\lambda}{\mu}} \sum_{k=K_0'}^{\infty} \left(\frac{\rho\lambda}{\mu}\right)^k / k! \tag{7.1}$$

where K'_0 satisfies the outage condition

$$\sum_{j=2}^{m} \nu_j < \frac{W / R(1-\eta)}{E_b / I_o} = K_0' \tag{7.2}$$

and ν_j is the binary random variable indicating whether the jth voice user is active at any instant. For example, for a process gain of 128, $\eta = 0.1$, and $E_b/N_0 = 5$, $K'_0 = 23$. If the voice activity factor is 1, the maximum number of users supported is $m = K'_0 + 1 = 24$.

Viterbi and Viterbi basically interpreted the outage probability as the blocking probability. However, the outage probability is not directly corresponding to the call blocking, as call blocking is mainly caused when a call is controlled by a CAC rule. That is, the call blocking and outage should be distinguished because the call blocking occurs when an incoming mobile cannot be admitted in the system, while the outage occurs when a mobile admitted in the cell cannot maintain the target QoS requirement.

In contrast with [8], we will in this chapter also extend the analysis of Erlang capacity to the case of multiclass CDMA systems, based on multidimension *M/M/m* loss model [9–11] (i.e., m server model with Poisson input and exponential service time such that when all m channels are busy, an arrival leaves the system without waiting for service, where *M* and *M* means that each user has exponentially distributed interarrival times and service times, and m means there is m finite number of available servers). More fundamental explanations on *M/M/m* queue are available in Appendix B.

The blocking probability with the *M/M/m* loss model is simply given by the Erlang B formula, rather than the Poisson distribution, but the Poisson distribution and the Erlang B formula practically arrive at the same results when the number of servers in the system is larger than 20 [9]. This approach also allows for the provision of different GoS for different types of calls. This is made possible by the introduction of a new GoS metric, the blocking probability in addition to the outage probability [11].

With this approach, the Erlang analysis of CDMA systems can be performed in two stages. In the first stage we determine the number of available, or *virtual*, trunks, called *trunk capacity*. In the second stage, we determine the Erlang capacity from the number of virtual trunks. The trunks are not physical trunks but rather virtual ones. Noting that the limitation of the underlying physical system is taken into account when evaluating the number of available trunks, we can refer to the trunking capacity as the maximum possible number of simultaneous users that can be supported by the system while the QoS requirements of each user (e.g., data rate, BER, and outage probability) are being satisfied. Figure 7.1 shows two stages to calculate the Erlang capacity, based on the multidimension *M/M/m* loss model.

The maximum allowable number of concurrent users that a CDMA system can support with QoS requirements has been found in many other papers [3, 5, 6],

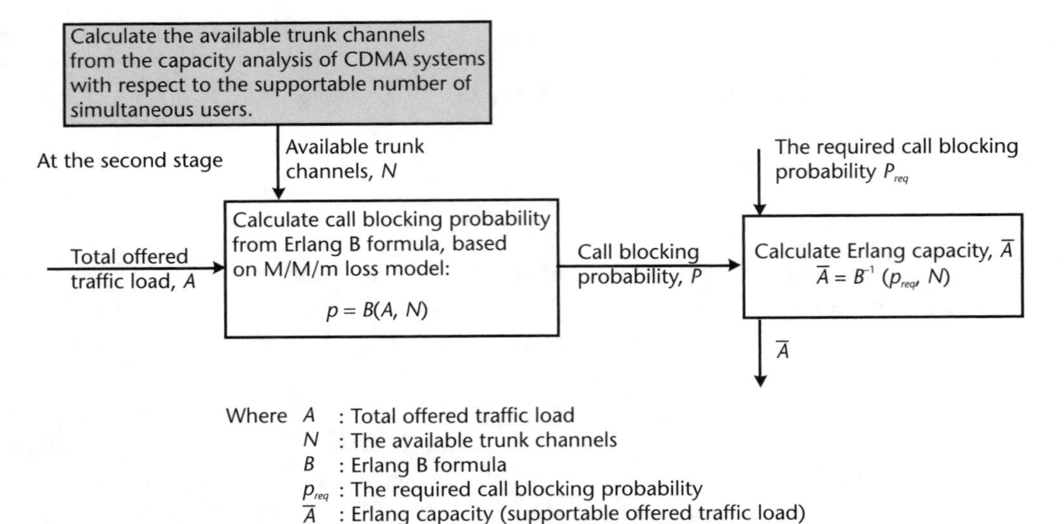

Figure 7.1 Two stages to calculate the Erlang capacity, based on the multidimension *M/M/m* loss model.

based on the maximum tolerable interference. In particular, as a result of [6], the system capacity limit of CDMA system supporting the K district service types (one voice and $K - 1$ data service groups) in the reverse link can be given as

$$\gamma_v n_v + \sum_{j=1}^{K-1} \gamma_{d_j} n_{d_j} \leq 1 \tag{7.3}$$

where

$$\gamma_v = \frac{\alpha}{\frac{W}{R_{v_{req}}}\left(\frac{E_b}{N_o}\right)_{v_{req}}^{-1}\left\langle\frac{1}{1+f}\right\rangle 10^{\frac{Q^{-1}(\beta)}{10}\sigma_x - 0.012\sigma_x^2} + \alpha}$$

$$\gamma_{d_j} = \frac{1}{\frac{W}{R_{d_{j,req}}}\left(\frac{E_b}{N_o}\right)_{d_{j,req}}^{-1}\left\langle\frac{1}{1+f}\right\rangle 10^{\frac{Q^{-1}(\beta)}{10}\sigma_x - 0.012\sigma_x^2} + 1}$$

All relevant parameters in these equations are defined and described in Section 3.1.

The inequality of (7.3) is the necessary and sufficient condition satisfying the system QoS requirements and indicates that calls of different types of services take different amount of system resources according to their QoS requirements (e.g., information data rate and the required bit energy-to-inference power spectral density ratio). In the following analysis, based on (7.3), we assume that one call attempt of data in the jth service group is equivalent to Λ_j call attempts of voice service, where Λ_j is defined as $\lfloor \gamma_{d_j} / \gamma_v \rfloor$ and $\lfloor x \rfloor$ denotes the greatest integer must be less than or equal to x. Then, (7.3) can be rewritten as follows:

$$n_v + \sum_{j=d_1}^{d_{K-1}} \Lambda_j \cdot n_j \leq \hat{C}_{ETC} \tag{7.4}$$

where $\hat{C}_{ETC} \equiv \lfloor 1 / \gamma_v \rfloor$ is the total number of basic channels, and subscript "ETC" denotes equivalent telephone (voice) channel. That is, the voice channel is presumed to the basic channel.

For safe network operation, it is of vital importance to define a suitable policy for the acceptance of an incoming call, in order to guarantee a certain QoS. In this chapter, a set of possible number of supportable users, which is limited by (7.3) or by (7.4), is defined as a call admission region for a CAC rule. In such a CAC rule, a call request is blocked and cleared from the system if its acceptance would move into the states out of the admissible region. Otherwise, a call request is accepted.

7.3 Erlang Capacity for the Multimedia CDMA Systems

We assume the system being considered is characterized as follows:

1. The calls of the jth service group in the home cell are generated as a Poisson process with arrival rate λ_j, and the arrival rate is homogeneous.
2. A call request is blocked and cleared from the system if its acceptance would move into the states out of the admissible region.
3. If a call is accepted, then it remains in the cell of its origin for a holding time that has an exponential distribution with the mean holing time $1/\mu_j$, where holding time is homogeneous and independent both of the other holding times and of the arrival processes.

Also, let us denote $(n_1, \ldots, n_K)$ as a state randomly selected to represent the number of concurrent users of a corresponding service group. With the previous assumptions, the system supporting K service groups can be modeled as a K-dimensional Markov chain. For example, Figure 7.2 depicts a state transition diagram in the case that a system supports two service groups (voice and data traffic), given the offered traffic loads.

According to the theory of circuit-switched networks [12], it is well known that there exists an equilibrium probability, $\pi(\mathbf{N})$ for an admissible state $\mathbf{N}(n_1, \ldots, n_K)$, and it is given by:

$$\pi(\mathbf{N}) = \frac{1}{G(R)} \prod_{i=1}^{K} \frac{\rho_i^{n_i}}{n_i!} \quad \text{for } \mathbf{N} \in S(R) \tag{7.5}$$

where $\rho_i = \lambda_i / \mu_i$, which denotes the offered traffic load of the ith service group.

$G(R)$ is a normalizing constant that has to be calculated in order to have the $\pi(\mathbf{N})$ that is accumulated to 1:

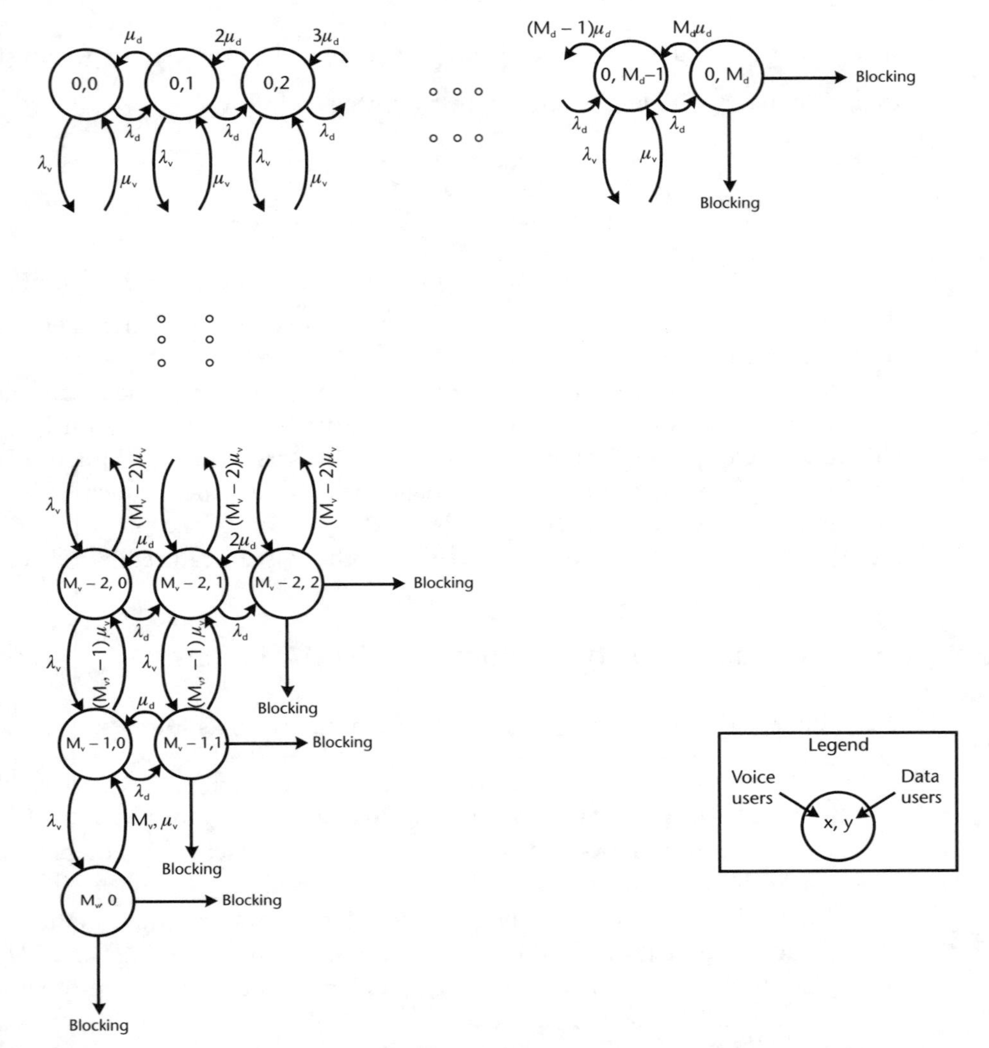

Figure 7.2 The state transition diagram for a CDMA system supporting voice and data services.

$$G(R) = \sum_{\mathrm{N} \in S(R)} \prod_{i=1}^{K} \frac{\rho_i^{n_i}}{n_i!} \tag{7.6}$$

For a multimedia CDMA system supporting K service groups, as we described in the previous section, a set of all admissible states can be given as:

$$S(R) = \{\mathbf{N}: \mathbf{N}\mathbf{A}^T \leq R\} \tag{7.7}$$

where $\mathbf{N}$ and $\mathbf{A}$ are 1 by K vector, respectively, and R is a scalar representing the system resource such that

$$\mathbf{A} = \left(1, \Lambda_{d_1}, \ldots, \Lambda_{d_{K-1}}\right) \text{ and } R = \hat{C}_{ETC} \tag{7.8}$$

Then, the call blocking probability, B_i of the ith service group, can be easily evaluated by means of two normalizing constants.

$$B_i = 1 - \frac{G(R - \mathbf{A}e_i)}{G(R)} \tag{7.9}$$

where e_i is a unit vector in the ith direction, and $G(R)$ is the normalizing constant calculated on the whole $S(R)$, while $G(R - \mathbf{A}e_i)$ is the normalizing constant calculated on the $S(R - \mathbf{A}e_i)$ with respect to the traffic of the ith service group.

In this situation, the Erlang capacity with respect to the ith service group can be calculated as a function of offered traffic loads of all service groups by contouring (7.9) at the required call blocking probability of the ith service group. This is because the Erlang capacity can be defined as a set of supportable offered traffic loads with a given quality and with availability of service as measured by the call blocking probability.

In order to consider all requirements of each service group, total system Erlang capacity, in this chapter, is defined as a set of offered traffic loads of all service groups in which all requirements of each service group are satisfied simultaneously. An easy way to visualize total system Erlang capacity is to consider the overlapped Erlang capacity region as total system Erlang capacity. If the system supports K service groups, then total system Erlang capacity is determined by the overlapped region of Erlang capacities with respect to the required call blocking probability of each service group whose dimension is determined by the number of service groups. Conceptually, it is expected that the Erlang capacities limited by the required call blocking probability of each service group should be balanced to get more large Erlang capacity.

7.4 Numerical Example

As a numerical example, let's consider a typical IS-95 CDMA system that supports voice and data services. The system parameters are shown in Table 7.1.

Figure 7.3 shows a two-dimensional system capacity bound with respect to the number of supportable users. All points (n_v, n_d) under the capacity plane represent a set of the possible number of concurrent users in the voice and data service groups, where n_v and n_d are integers. As aforementioned, a set of the possible user numbers under the capacity plane is used as the call admission region for the CAC rule. With the CAC rule and the given offered traffic loads of voice and data calls, the system state transition diagram is depicted in Figure 7.2, where $M_d = 6$, $M_v = 28$. The call blocking probabilities experienced by voice and data calls can be calculated as a function of the offered traffic loads using (7.9).

The corresponding call blocking probabilities of voice and data calls are depicted in Figures 7.4 and 7.5, respectively. Figure 7.6 shows the Erlang capacity region that the system can support when the required call blocking probabilities for voice and data traffic are given 5% and 1%, respectively. In Figure 7.6, the dashed line and solid line indicate the Erlang capacity bounds that are limited by the required call blocking probability of voice and data traffic, respectively. From

Table 7.1 System Parameters for the Numerical Example

Parameters	*Symbol*	*Value*
Allocated frequency bandwidth	W	1.25 Mbps
Required bit transmission rate for voice traffic	R_v	9.6 Kbps
Required bit transmission rate for data traffic	R_d	8 Kbps
Required bit energy-to-interference power spectral density ratio for voice traffic	$\left(\frac{E_b}{N_o}\right)_{v\,req}$	7 dB
Required bit energy-to-interference power spectral density ratio for data traffic	$\left(\frac{E_b}{N_o}\right)_{d\,req}$	10 dB
System reliability requirement	$\beta\%$	99%
Frequency reuse factor	$\left\langle\frac{1}{1+f}\right\rangle$	0.7
Standard deviation of received SIR	σ_x	1 dB
Activity factor for voice	σ_v	3/8
Activity factor for data	σ_d	1

Figure 7.6, two main facts are observed. The first fact is that data users have more impact than voice users on the Erlang capacity because the effective bandwidth of one data user is larger than that of one voice user in the numerical example. The other fact is that the total system Erlang capacity region that the system can support is determined not by the Erlang capacity limited by the call blocking probability of voice calls but by that of data calls, as the system should satisfy the required call blocking probabilities of voice and data calls, simultaneously. As predicted previously, it is required that Erlang capacities limited by the required call blocking probability of two service groups should be balanced to enhance total system Erlang capacity. For this purpose, some resource management schemes should be considered. In this chapter, we consider the channel reservation scheme in which some channels are reserved for certain service groups and the remaining channels are

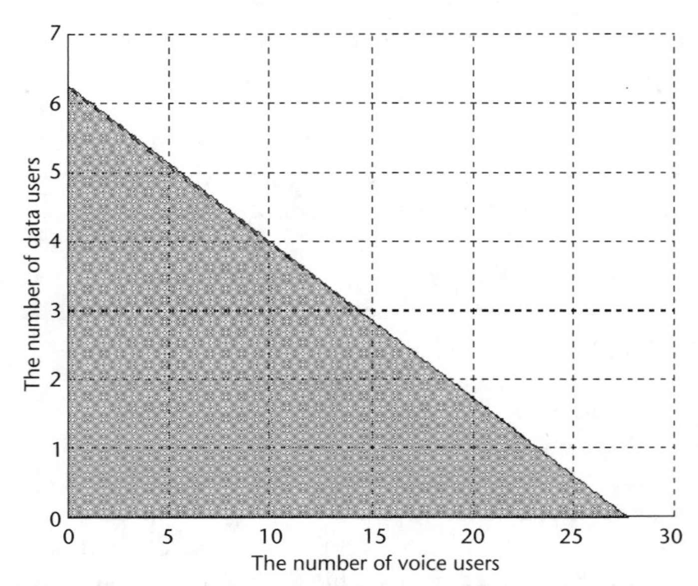

Figure 7.3 Capacity plane for two service groups with respect to the number of supportable users.

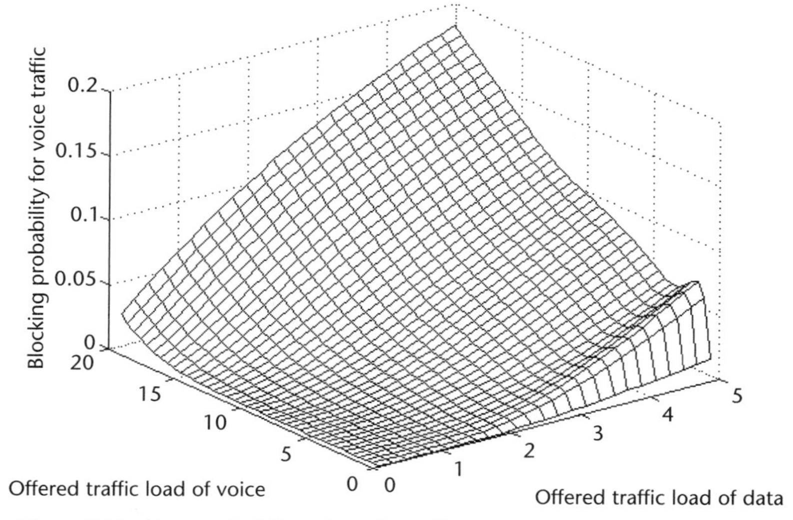

Figure 7.4 The call blocking probability of a voice call.

allocated to all service groups. That is, if we assume that χ channels are reserved for the *i*th service group, the users of the *i*th service group will only be accepted when there are less than χ channels in the system. Figure 7.7 depicts the state transition diagram for two service cases when one channel is reserved for data calls.

In the case of the numerical example, some channels should be reserved for data service because the Erlang capacity limited by the required call blocking probability of data calls is smaller than that of voice call. Figure 7.8 shows the effect of the reservation scheme on total system Erlang capacity when the required call blocking probabilities of voice and data calls are given as 5% and 1%, respectively.

As we can see in Figure 7.8, the Erlang capacity region limited by the required call blocking probability of data calls increases more than that of Figure 7.6 as the

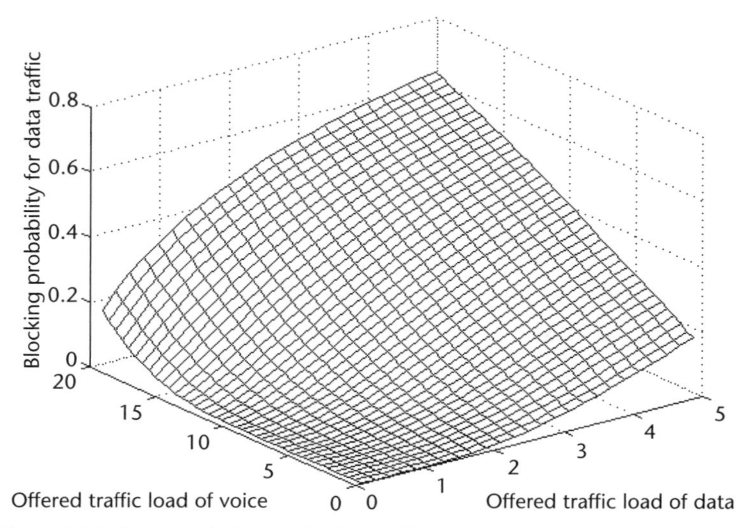

Figure 7.5 The call blocking probability of a data call.

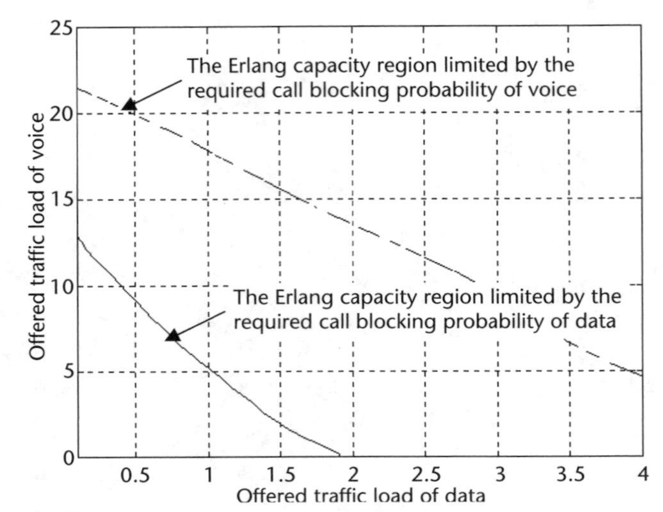

Figure 7.6 The Erlang capacity when at least 5% and 1% call blocking probability is needed for voice and data calls, respectively.

number of reservation channels for data calls increases, especially at the high offered traffic load of voice. On the other hand, the Erlang capacity region limited by the required call blocking probability of voice call decreases. However, we can observe that total system Erlang capacity is more or less increased by reserving two channels for data calls than that without the channel reservation scheme, by comparing Figures 7.8(a) and 7.8(c). In particular, the Erlang capacity region marked by the circle in Figure 7.8(c) indicates the amount of Erlang capacity that is improved through the reservation scheme. However, the amount of Erlang capacity improvement through the reservation scheme is not as large as expected. Hence, some other resource management schemes should be suggested to make the Erlang capacities with respect to each service group more efficiently balanced.

Even though a CDMA system supporting voice and data services has been considered so far, it should be noteworthy that the proposed method can be applied to calculate the Erlang capacity of CDMA systems supporting various service types.

In addition, to include the soft handover mechanism in the Erlang capacity analysis, which is a key technology in realizing CDMA cellular system, we may take following analysis procedures. First, we need to characterize the features of soft handover calls, such as a channel holding time, and classify the traffic into new and handover calls according to their traffic characteristics. Then, two service groups (voice and data calls) will be expanded to four service groups (new voice, new data, handover voice, and handover data calls) for the cases of numerical examples. In particular, the effect of handoff calls on Erlang capacity is investigated in Chapter 6. Finally, Erlang capacity can be found in four-dimensional observation space by using a procedure similar to that presented in this chapter.

7.5 Conclusion

In this chapter, we have presented an analytical approach for evaluating the Erlang capacity of multimedia CDMA systems in the reverse link, based on a multi-

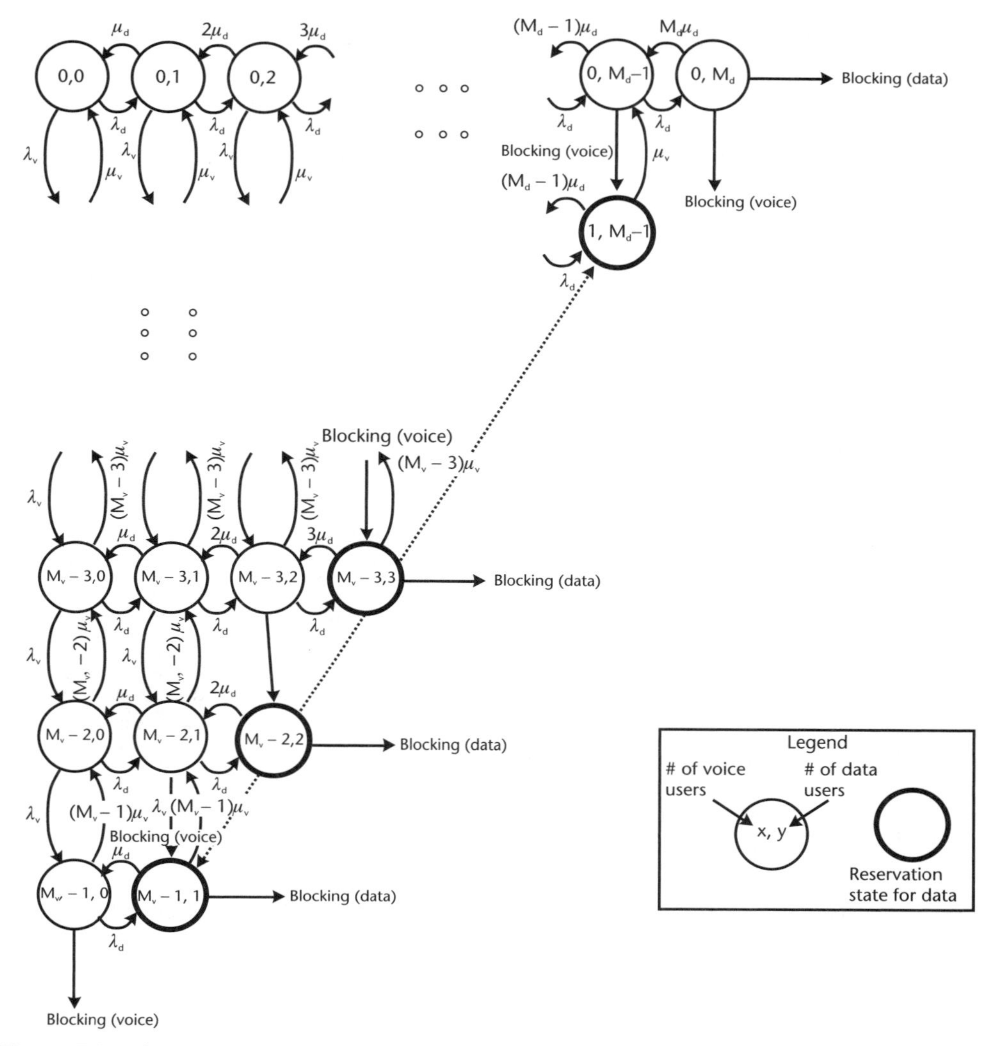

Figure 7.7 The state transition diagram when the reservation scheme is used and one channel is exclusively reserved for data calls.

dimensional *M/M/m* loss model, where the capacity bound with respect to the maximum number of supportable users is utilized as a reference for the CAC rule. Through a numerical example, we observe that data users have more impact on the Erlang capacity than do voice users, as the effective bandwidth of one data call is larger than that of one voice call. It is also necessary to find a balance between the Erlang capacities with respect to each service group to enhance total system Erlang capacity. As a solution, the channel reservation scheme is considered, and it is also observed that total system Erlang capacity can be increased by properly reserving some channels for prioritized calls. In the channel reservation scheme that has been considered so far, fixed reservation channels are exclusively allocated for prioritized calls without any reference to the offered traffic load. This kind of fixed reservation scheme may result in the inefficiency of system resource utilization, especially at a low traffic load of prioritized calls. Hence, it is a remaining work to observe the effect of dynamic reservation schemes on the Erlang capacity, where we allocate

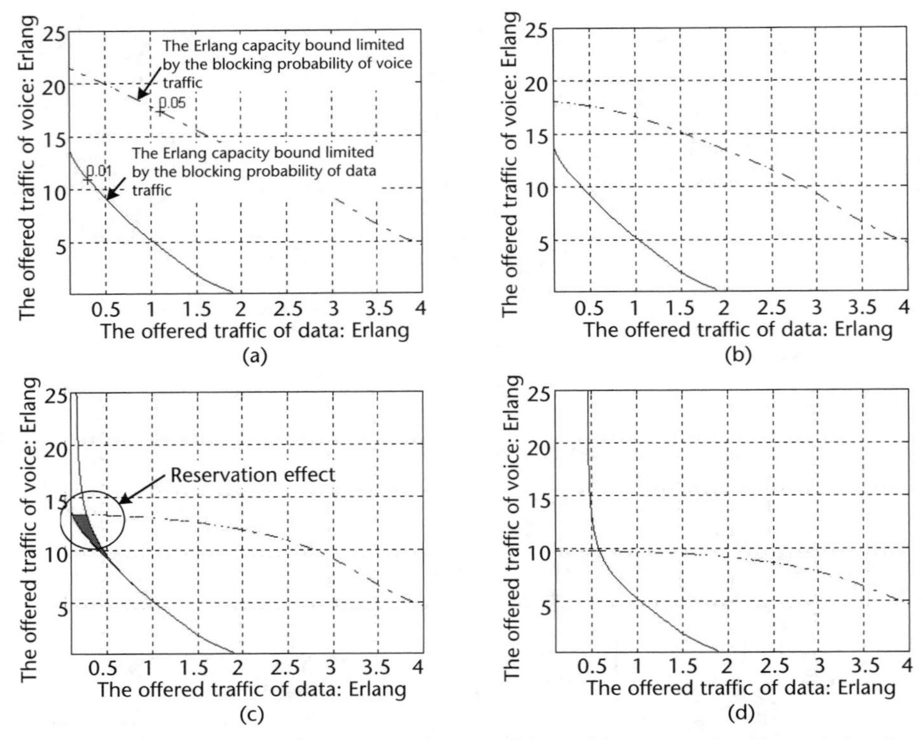

Figure 7.8 The Erlang capacity according to the number of the reservation channels for data calls: (a) when there is no reservation channel, (b) when the number of reservation channels for data calls is one, (c) when the number of reservation channels for data calls is two, and (d) when the number of reservation channels for data calls is three.

reservation channels dynamically for prioritized calls by considering the amount of the offered traffic load.

References

[1] Kim, K. I., *Handbook of CDMA System Design, Engineering and Optimization*, Englewood Cliffs, NJ: Prentice Hall, 2000.

[2] Wu, J. S., and J. R. Lin, "Performance Analysis of Voice/Data Integrated CDMA System with QoS Constraints," *IEICE Trans. on Communications*, Vol. E79-B, 1996, pp. 384–391.

[3] Sampath, A., P. S. Kumar, and J. M. Holtzman, "Power Control and Resource Management for a Multimedia CDMA Wireless System," *IEEE Proc. of International Symposium on Personal, Indoor, and Mobile Radio Communications*, 1995, pp. 21–25.

[4] Sasaki, A., et al., "Standardization Activities on FPLMTS Radio Transmission Technology in Japan," *IEICE Trans. Fundamentals*, 1996, pp. 1938–1946.

[5] Yang, J. R., et al., "Capacity Plane of CDMA System for Multimedia Traffic," *IEE Electronics Letters*, 1997, pp. 1432–1433.

[6] Koo, I., et al., "A Generalized Capacity Formula for the Multimedia DS-CDMA System," *IEEE Proc. of Asia-Pacific Conference on Communications*, 1997, pp. 46–50.

[7] Sampath, A., N. B. Mandayam, and J. M. Holtzman, "Erlang Capacity of a Power Controlled Integrated Voice and Data CDMA System," *IEEE Proc. of Vehicular Technology Conference*, 1997, pp. 1557–1561.

[8] Viterbi, A. M., and A. J. Viterbi, "Erlang Capacity of a Power-Controlled CDMA System," *IEEE Journal on Selected Areas in Communications,* 1993, pp. 892–900.

[9] Jacobsmeyer, J., "Congestion Relief on Power-Controlled CDMA Networks," *Selected IEEE Journal on Areas in Communications,* 1996, pp. 1758–1761.

[10] Koo, I., et al., "Analysis of Erlang Capacity for the Multimedia DS-CDMA System," *IEICE Trans. Fundamentals,* 1999, pp. 849–855.

[11] Matragi, W., and S. Nanda, "Capacity Analysis of an Integrated Voice and Data CDMA System," *IEEE Proc. of Vehicular Technology Conference,* 1999, pp. 1658–1663.

[12] Kelly, F., "Loss Networks," *The Annals of Applied Probability*, 1991, pp. 319–378.

CHAPTER 8

Erlang Capacity Under the Delay Constraint

Drs. J. Yang and K. Kim

In this chapter, we analyze the Erlang capacity of a CDMA system supporting voice and delay-tolerant data services and consider the characteristics of delay-tolerant traffic, known as the *delay confidence*. Delay confidence is defined as the probability that a new data call is accepted within the maximum tolerable delay without being blocked. In this case, the Erlang capacity is confined not only by the required blocking probability of voice call but also by the required delay confidence of data call. For the performance analysis, we develop a two-dimensional Markov chain model, based on the first-come-first-served (FCFS) service discipline, and present a numerical procedure to analyze the Erlang capacity. As a result, it is necessary to create a balance between the Erlang capacity with respect to the blocking probability of voice calls and one with respect to the delay confidence of data calls, in order to accommodate more Erlang capacity. In this chapter, we demonstrate the balance by properly selecting the size of the designated queue for data traffic.

8.1 Introduction

The objective of future wireless communication systems is to provide users with multimedia services (e.g., voice, interactive data, file transfer, Internet access, and images) comparable to those provided by the wired communication systems.

Different traffic types may have different QoS requirements, which makes the capacity evaluation more complex. Many efforts have been made to analyze the capacity of a CDMA system. Typically, the capacity of a CDMA system has been defined as the maximum number of users or the Erlang capacity [1–4]. The former and latter definitions of the capacity are used for estimating a supportable size of the system at one time and for measuring the economic usefulness of the system, respectively [1]. In [1, 2], the outage probability was presumed to be the call blocking probability, and the call blocking probabilities of different traffic types in the system were represented identically. By using a multidimensional Markov loss model, based on the maximum number of supportable current users, the call blocking probabilities of different traffic types were considered separately, and the Erlang capacity was analyzed with respect to the required blocking probabilities of different traffic types [5, 6].

Voice and data traffic are generally considered delay intolerant and delay tolerant, respectively. To achieve higher capacity using the delay-tolerant characteristic of data traffic, data calls can be queued until the required resources are available in the system. The blocking probability and the average delay have been typically considered as a performance measure for delay-tolerant traffic [5, 7]. However, the more meaningful measurement for delay-tolerant traffic is the delay confidence rather than the average delay, where the delay confidence is defined as the probability that a new data call gets a service within the maximum tolerable delay requirement without being blocked. Noting that the previous works [1, 5–7] have not considered delay confidence when evaluating the Erlang capacity, in this chapter we adopt the delay confidence as a performance measure of delay-tolerant traffic and further analyze Erlang capacity of a CDMA system supporting voice and data traffic. Here, the Erlang capacity is defined as a set of average offered traffic loads of voice and data calls that can be supported in the system while the required blocking probability of voice calls and the required delay confidence of data calls are being satisfied simultaneously. To analyze the Erlang capacity, we develop a two-dimensional Markov chain model, based on the FCFS service discipline, where a queue with finite size is exclusively designated for delay-tolerant data calls. Based on the Markov chain model, we present a numerical procedure to analyze the call blocking probability of voice and data calls, the delay distribution, and delay confidence of data calls, all of which are necessary to analyze the Erlang capacity. In addition, a procedure selecting the proper size of the queue length for data traffic is suggested in order to accommodate more Erlang capacity in the system.

The remaining chapter is organized as follows. In the next section, a CAC scheme is stipulated based on system capacity in terms of the maximum number of supportable users. In Section 8.3, we develop a two-dimensional Markov chain model and analyze the blocking probabilities of voice and data calls. Based on the Markov chain model, Section 8.4 shows an analytical approach to evaluating the delay distribution of data calls. With the blocking probability and delay distribution, we analyze the delay confidence in Section 8.5. In Section 8.6, the Erlang capacity is analyzed, which can be supported in the system while the required blocking probability of voice traffic and the required delay confidence of data traffic are being satisfied simultaneously. Finally, conclusions are remarked in Section 8.7.

8.2 System Model

In CDMA systems, although there is no hard limit on the number of concurrent users, there is a practical limit on the number of concurrent users in order to control the interference among users that share the same pilot signal; otherwise, the system can fall into an outage state where QoS requirements of users cannot be guaranteed. In order to satisfy the QoS requirements of all concurrent users, the capacity of CDMA systems supporting voice and data services in the reverse link should be limited with following equation [4]

$$\gamma_v i + \gamma_d j \le 1, \quad i \text{ and } j \ge 0 \tag{8.1}$$

where

$$\gamma_v = \left(\frac{W}{R_v q_v} + 1\right)^{-1} \text{ and } \gamma_d = \left(\frac{W}{R_d q_d} + 1\right)^{-1} \tag{8.2}$$

γ_v and γ_d are the amount of system resources that are used by one voice and one data user, respectively. i and j denote the number of users in the voice and data service groups, respectively. W is the allocated frequency bandwidth. q_v and q_d are the bit energy-to-interference power spectral density ratio for voice and data calls, respectively, which is required to achieve the target BER at the BS. R_v and R_d are the required information data rates of voice and data service groups, respectively. Each user is classified by his or her own QoS requirements, such as the required information data rate and the required bit energy-to-interference spectral density ratio, and all users in same service group have the same QoS requirements.

Equation (8.1) indicates that the calls that have different types of services take different amounts of system resources according to their QoS requirements.

We also assume that the system employs a circuit switching method to handle the transmission of voice and data calls. Each call shares the system resources with the other calls, and they contend for the use of system resources. Once a call request is accepted in the system, the call occupies the required amount of system resources and transmits the information without any delay throughout the call duration.

With regard to network operation, it is of vital importance to set up a suitable policy for the acceptance of an incoming call in order to guarantee a certain QoS. In general, CAC policies can be divided into two categories: NCAC and ICAC [8]. NCAC implies that a call will or won't be accepted, depending on whether the number of concurrent users is greater than a threshold. In the case of ICAC, a BS determines whether a new call is acceptable by monitoring the interference level on a call-by-call basis, while the NCAC utilizes a predetermined CAC threshold. In this chapter, we adopt an NCAC-type CAC due to its simplicity, even though the NCAC generally suffers a slight performance degradation over the ICAC [8]. We also adopt the capacity bound, stipulated by (8.1) as a predetermined CAC threshold. Further, we consider the queue with the finite length of K for delay-tolerant data traffic to exploit its delay-tolerant characteristic, and we use the FCFS rule as a service discipline. Based these assumptions, the CAC rule, for the case $\gamma_d > \gamma_v$ can be summarized as follows:

- If $\gamma_v i + \gamma_d j \le 1 - \gamma_d$, then both new voice and new data calls are accepted.
- If $1 - \gamma_d < \gamma_v i + \gamma_d j \le 1 - \gamma_v$, then new voice calls are accepted, and new data calls are queued.
- If $1 - \gamma_d < \gamma_v i + \gamma_d j \le 1 + (K - 1)\gamma_d$, then new voice calls are blocked, and new data calls are queued.
- If $\gamma_v i + \gamma_d j > 1 + (K - 1)\gamma_d$, then both new voice and new data calls are blocked.

Here, we set one voice channel as the basic channel. In this case, the number of total basic channels in the system is $1/\gamma_v$, and the number of basic channels required by one data call is given as γ_d/γ_v, respectively. Here, it is noteworthy that the number of total basic channels in the system and the number of basic channels required by one data call are integer numbers in TDMA or FDMA, whereas they are real numbers in CDMA systems [6].

In order to analyze the performance of the system under the CAC policy, the arrivals of voice and data calls are assumed to be distributed according to independent Poisson processes with the average arrival rate λ_v and λ_d, respectively.

The service times of voice and data calls are assumed to be exponentially distributed with the average service time $1/\mu_v$ and $1/\mu_d$, respectively. Then, the offered traffic loads of voice and data calls are expressed as $\rho_v = \lambda_v/\mu_v$ and $\rho_d = \lambda_d/\mu_d$, respectively.

8.3 Markov Chain Model and Blocking Probability

In this section, we develop an analytical model to determine the blocking probabilities of voice and data calls. The model will also be utilized to analyze the delay distribution of data call in the next section.

According to the CAC rule based on the number of concurrent users, the set of possible admissible states is given as

$$\Omega_S = \left\{(i,j) | 0 \le i \le \gamma_v^{-1}, j \ge 0, \gamma_v i + \gamma_d j \le 1 + \gamma_d K\right\} \tag{8.3}$$

For these admissible states, Figure 8.1 shows five distinct regions and a typical state transition for each region to represent the call-level state transition diagram. These possibly admissible states divided into five regions are as follows:

$$\begin{aligned}
\Omega_A &= \left\{(i,j) | 0 \le \gamma_v \cdot i + \gamma_d \cdot j \le 1 - \max(\gamma_v, \gamma_d)\right\} \\
\Omega_B &= \left\{(i,j) | 1 - \max(\gamma_v, \gamma_d) < \gamma_v \cdot i + \gamma_d \cdot j \le 1 - \min(\gamma_v, \gamma_d)\right\} \\
\Omega_C &= \left\{(i,j) | 1 - \min(\gamma_v, \gamma_d) < \gamma_v \cdot i + \gamma_d \cdot j \le 1\right\} \\
\Omega_D &= \left\{(i,j) | 1 < \gamma_v \cdot i + \gamma_d \cdot j \le 1 + \gamma_d \cdot (K-1)\right\} \\
\Omega_E &= \left\{(i,j) | 1 + \gamma_d \cdot (K-1) < \gamma_v \cdot i + \gamma_d \cdot j \le 1 + \gamma_d \cdot K\right\}
\end{aligned} \tag{8.4}$$

Noting that total rate of flowing into a state (i, j) is equal to that of flowing out, we can get the steady-state balance equation for each state as follows:

$$\begin{aligned}
&\text{Rate-In} = \text{Rate-Out} \\
&\text{Rate-In} = \vec{a} \cdot P_{i+1,j} + \vec{b} \cdot P_{i,j+1} + \vec{c} \cdot P_{i-1,j} + \vec{d} \cdot P_{i,j-1} \\
&\text{Rate-Out} = \left(\vec{i} + \vec{j} + \vec{k} + \vec{l}\right) \cdot P_{i,j} \\
&\qquad\qquad \text{for all states}
\end{aligned} \tag{8.5}$$

where the state transition rates, $\vec{a}, \vec{b}, \vec{c}, \vec{d}, \vec{e}, \vec{f}, \vec{g}, \vec{h}, \vec{i}, \vec{j}, \vec{k}$, and $\vec{l}$ involved in (8.5) can be given by as follows:

$$\begin{aligned}
\vec{a} &\equiv \text{transition rate from state } (i+1, j) \text{ to state } (i, j) \\
&= \begin{cases} (i+1)\mu_v & (i,j) \in \Omega_S \\ 0 & \text{otherwise} \end{cases}
\end{aligned} \tag{8.6}$$

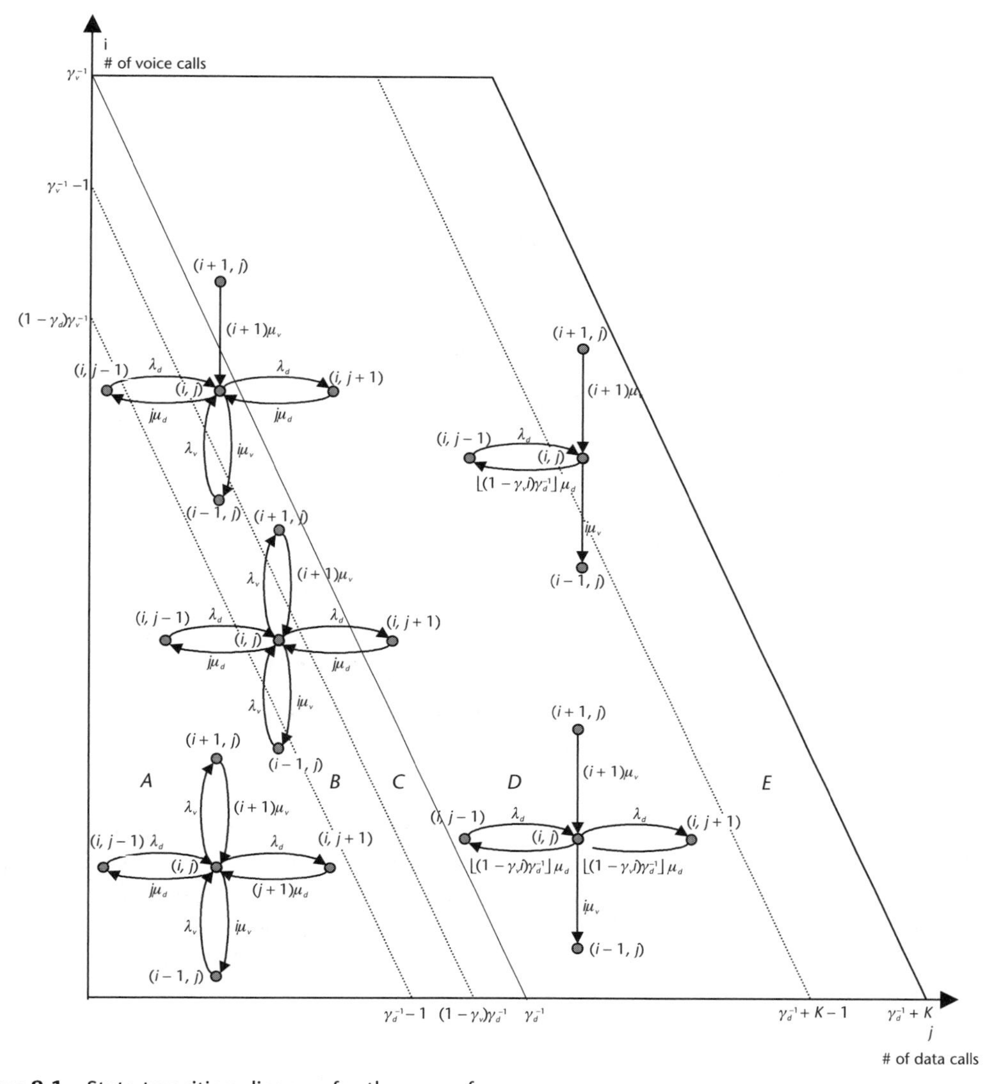

Figure 8.1 State transition diagram for the case of $\gamma_d > \gamma_v$.

$$\vec{b} \equiv \text{transition rate from state } (i, j+1) \text{ to state } (i, j)$$

$$= \begin{cases} (i+1)\mu_d & (i, j) \in \Omega_A \\ j \cdot \mu_d & (i, j) \in \{\Omega_B, \Omega_C\} \\ \lfloor (1-\gamma_v \cdot i)\gamma_d^{-1} \rfloor \cdot \mu_d & (i, j) \in \Omega_D \\ 0 & \text{otherwise} \end{cases} \tag{8.7}$$

$$\vec{c} \equiv \text{transition rate from state } (i-1, j) \text{ to state } (i, j)$$

$$= \begin{cases} \lambda_v & (i, j) \in \{\Omega_A, \Omega_B, \Omega_C\} \\ 0 & \text{otherwise} \end{cases} \tag{8.8}$$

$$\vec{d} \equiv \text{transition rate from state } (i, j-1) \text{ to state } (i, j) = \begin{cases} \lambda_d & (i,j) \in \Omega_S \\ 0 & \text{otherwise} \end{cases} \tag{8.9}$$

$$\vec{i} \equiv \text{transition rate from state } (i, j) \text{ to state } (i+1, j) = \begin{cases} \lambda_v & (i,j) \in \{\Omega_A, \Omega_B\} \\ 0 & \text{otherwise} \end{cases} \tag{8.10}$$

$$\vec{j} \equiv \text{transition rate from state } (i, j) \text{ to state } (i, j+1) = \begin{cases} \lambda_d & (i,j) \in \{\Omega_A, \Omega_B, \Omega_C, \Omega_D\} \\ 0 & \text{otherwise} \end{cases} \tag{8.11}$$

$$\vec{k} \equiv \text{transition rate from state } (i, j) \text{ to state } (i-1, j) = \begin{cases} i\mu_v & (i,j) \in \Omega_S \\ 0 & \text{otherwise} \end{cases} \tag{8.12}$$

$$\vec{l} \equiv \text{transition rate from state } (i, j) \text{ to state } (i, j-1) = \begin{cases} i\mu_d & (i,j) \in \{\Omega_A, \Omega_B, \Omega_C\} \\ \lfloor (1-\gamma_v \cdot i)\gamma_d^{-1} \rfloor & \text{otherwise} \end{cases} \tag{8.13}$$

Figure 8.2 summarizes the steady-state balance equations for the state transit diagram according to the region to which the current state belongs. If the total number of all possible states is n_s, the balance equations become $(n_s - 1)$ linearly independent equations. With these $(n_s - 1)$ equations and the normalized equation $\sum_{(i,j)\in\Omega_S} P_{i,j} = 1$, a set of n_s linearly independent equations for the state diagram can be formed as

$$\mathbf{Q}\pi = \mathbf{P} \tag{8.14}$$

where $\mathbf{Q}$ is the coefficient matrix of the n_s linear equations, π is the vector of state probabilities, and $\mathbf{P} = [0, \ldots, 0, 1]^T$. The dimensions of $\mathbf{Q}$, π, and $\mathbf{P}$ are $n_s \times n_s$, $n_s \times 1$, and $n_s \times 1$, respectively. By solving $= \mathbf{Q}^{-1}\mathbf{P}$, we can obtain the steady-state probabilities of all states [5].

Based on the CAC rule, a new voice call will be blocked if the channel resources are not enough to accept the call, and the corresponding blocking probability for voice calls is given by

$$P_{b_v} = \sum_{(i,j)\in\Omega_{(nv,blo)}} P_{i,j} \tag{8.15}$$

where

$$\Omega_{(nv,blo)} = \{(i,j) | \gamma_v i + \gamma_d j > 1 - \gamma_v\} \tag{8.16}$$

A:

$$(\lambda_v + \lambda_d + i\mu_v + j\mu_d)P_{i,j} = \lambda_v P_{i-1,j} + \lambda_d P_{i,j-1} + (i+1)\mu_v P_{i+1,j} + (j+1)\mu_d P_{i,j+1}$$

$$\text{for } (i,j) \in \left\{(i,j) \,\middle|\, 0 \le (i,j)\begin{pmatrix}\gamma_v\\ \gamma_d\end{pmatrix} \le 1 - \max\{\gamma_v, \gamma_d\}\right\}$$

⇨ Both new voice and new data calls are accepted

If $\gamma_v < \gamma_d$,

B:

$$(\lambda_v + \lambda_d + i\mu_v + j\mu_d)P_{i,j} = \lambda_v P_{i-1,j} + \lambda_d P_{i,j-1} + (i+1)\mu_v P_{i+1,j} + j\mu_d P_{i,j+1}$$

$$\text{for } (i,j) \in \left\{(i,j) \,\middle|\, 1 - \gamma_d < (i,j)\begin{pmatrix}\gamma_v\\ \gamma_d\end{pmatrix} \le 1 - \gamma_v\right\}$$

⇨ New voice calls are accepted and new data calls are queried.

If $\gamma_v > \gamma_d$,

$$(\lambda_d + i\mu_v + j\mu_d)P_{i,j} = \lambda_v P_{i-1,j} + \lambda_d P_{i,j-1} + (i+1)\mu_v P_{i+1,j} + (j+1)\, j\mu_d P_{i,j+1}$$

$$\text{for } (i,j) \in \left\{(i,j) \,\middle|\, 1 - \gamma_v < (i,j)\begin{pmatrix}\gamma_v\\ \gamma_d\end{pmatrix} \le 1 - \gamma_d\right\}$$

⇨ New voice calls are blocked and new data calls are accepted.

C:

$$(\lambda_d + i\mu_v + j\mu_d)P_{i,j} = \lambda_v P_{i-1,j} + \lambda_d P_{i,j-1} + (i+1)\mu_v P_{i+1,j} + j\mu_d P_{i,j+1}$$

$$\text{for } (i,j) \in \left\{(i,j) \,\middle|\, 1 - \min\{\gamma_v, \gamma_d < (i,j)\begin{pmatrix}\gamma_v\\ \gamma_d\end{pmatrix} \le 1\right\}$$

⇨ New voice calls are blocked and new data calls are queued.

D:

$$(\lambda_d + i\mu_v + \lfloor(1 - \gamma_v i)\gamma_d^{-1}\rfloor \mu_d)P_{i,j} = \lambda_d P_{i,j-1} + (i+1)\mu_v P_{i+1,j} + \lfloor(1 - \gamma_d i)\gamma_d^{-1}\rfloor \mu_d P_{i,j+1}$$

$$\text{for } (i,j) \in \left\{(i,j) \,\middle|\, 1 < (i,j)\begin{pmatrix}\gamma_v\\ \gamma_d\end{pmatrix} \le 1 + \gamma_d(K-1)\right\}$$

⇨ New voice calls are blocked and new data calls are queued.

E:

$$(i\mu_v + \lfloor(1 - \gamma_v i)\gamma_d^{-1}\rfloor \mu_d)P_{i,j} = \lambda_d P_{i,j-1} + (i+1)\mu_v P_{i+1,j}$$

$$\text{for } (i,j) \in \left\{(i,j) \,\middle|\, 1 + \gamma_d(K-1) < (i,j)\begin{pmatrix}\gamma_v\\ \gamma_d\end{pmatrix} \le 1 + \gamma_d K\right\}$$

⇨ Both new voice and new data calls are blocked.

Figure 8.2 Steady-state balance equations corresponding to the voice/data CDMA system.

$\Omega_{(nv,blo)}$ is composed of the regions C, D, and E in Figure 8.1. Similarly, a new data call will be blocked if the queue is full, and the blocking probability for data calls is given by

$$P_{b_d} = \sum_{(i,j)\in\Omega_{(nd,blo)}} P_{i,j} \tag{8.17}$$

where

$$\Omega_{(nd,blo)} = \{(i,j) | \gamma_v i + \gamma_d i > 1 + \gamma_d (K-1)\} \tag{8.18}$$

$\Omega_{(nd,blo)}$ corresponds to region E in Figure 8.1.

8.4 Delay Distribution

In this section, for this purpose, we will derive the cumulative distribution function (CDF) of delay and the delay confidence of data traffic. First, let's derive the CDF of delay (τ), based on the Markov chain model depicted in Figure 8.1. The delay is defined as the time that a data call waits in a queue until it is accepted in the system. For the convenience of analysis, we separate the CDF of delay into two parts corresponding to discrete and continuous parts of the random variable τ. That is,

$$F_d(t) = \Pr\{\tau \le t\} = F_d(0) + G(t) \tag{8.19}$$

where $F_d(0) = \Pr\{\tau \le 0\}$, and $G(t)$ represents the continuous part of the delay.

At first, the discrete part $F_d(0)$, represents the case when the delay is zero, and it can be calculated as follows:

$$\begin{aligned} F_d(0) &= \Pr\{\tau \le 0\} = \Pr\{\tau = 0\} \\ &= \sum_{(i,j)\in\Omega_{(nd,acc)}} P'_{i,j} \end{aligned} \tag{8.20}$$

where $\Omega_{(nd,acc)}$ is the acceptance region of new data calls, which is given as

$$\Omega_{(nd,acc)} = \{(i,j) | \gamma_v i + \gamma_d j \le 1 - \gamma_d\} \tag{8.21}$$

and

$$P'_{i,j} = \frac{P_{i,j}}{1 - P_{b_d}} \tag{8.22}$$

$P'_{i,j}$ represents the probability that there are i voice and j data calls in the system just before a new data call is admitted. If the state (i, j) belongs to the blocking region of new data calls, $\Omega_{(nd,blo)}$, the call will be blocked.

To investigate the continuous part of delay $G(t)$, let (i', j') denote the number of calls, excluding the number of service-completed calls within time τ from (i, j).

Consider the case that (i, j) belongs to the queuing region of new data calls just before a new data call is admitted, where the queuing region of new data calls is given as

$$\Omega_{(nd,que)} = \{(i,j) | 1 - \gamma_d < \gamma_v i + \gamma_d j \le 1 + (K-1)\gamma_d\} \tag{8.23}$$

In order for a new data call to be accepted within the time t according to the FCFS service discipline, (i', j') should fall into the acceptance region of new data calls within the time t. $G(t)$ is the sum of the probabilities of all cases that a state (i, j) in $\Omega_{(nd,que)}$ changes into (i', j') in $\Omega_{(nd,acc)}$ within the time, t. This can be expressed as

$$G(t) = \sum_{(i,j)\in\Omega_{(nd,que)}} \Pr\left\{\begin{matrix}(i',j')\in\Omega_{(nd,acc)}\text{within time } t| \\ \text{the system state is } (i,j) \text{ just before a new data call is admitted}\end{matrix}\right\}\cdot P'_{i,j}$$
$$= \sum_{(i,j)\in\Omega_{(nd,que)}} \int_0^t w_{(i,j)}(\tau)d\tau \cdot P'_{i,j} \tag{8.24}$$

where $w_{(i,j)}(\tau)$ is the delay distribution for the state (i, j), and it represents the probability that a new data call will be accepted within time τ, given that the system state is (i, j) just before the call is admitted. For example, Figure 8.3 shows the set of states representing the admissible numbers of voice and data calls for the case that W = 1.25 MHz, q_v = 7 dB, q_d = 7 dB, R_v = 9.6 Kbps, R_d = 19.2 Kbps, and K = 3. Consider

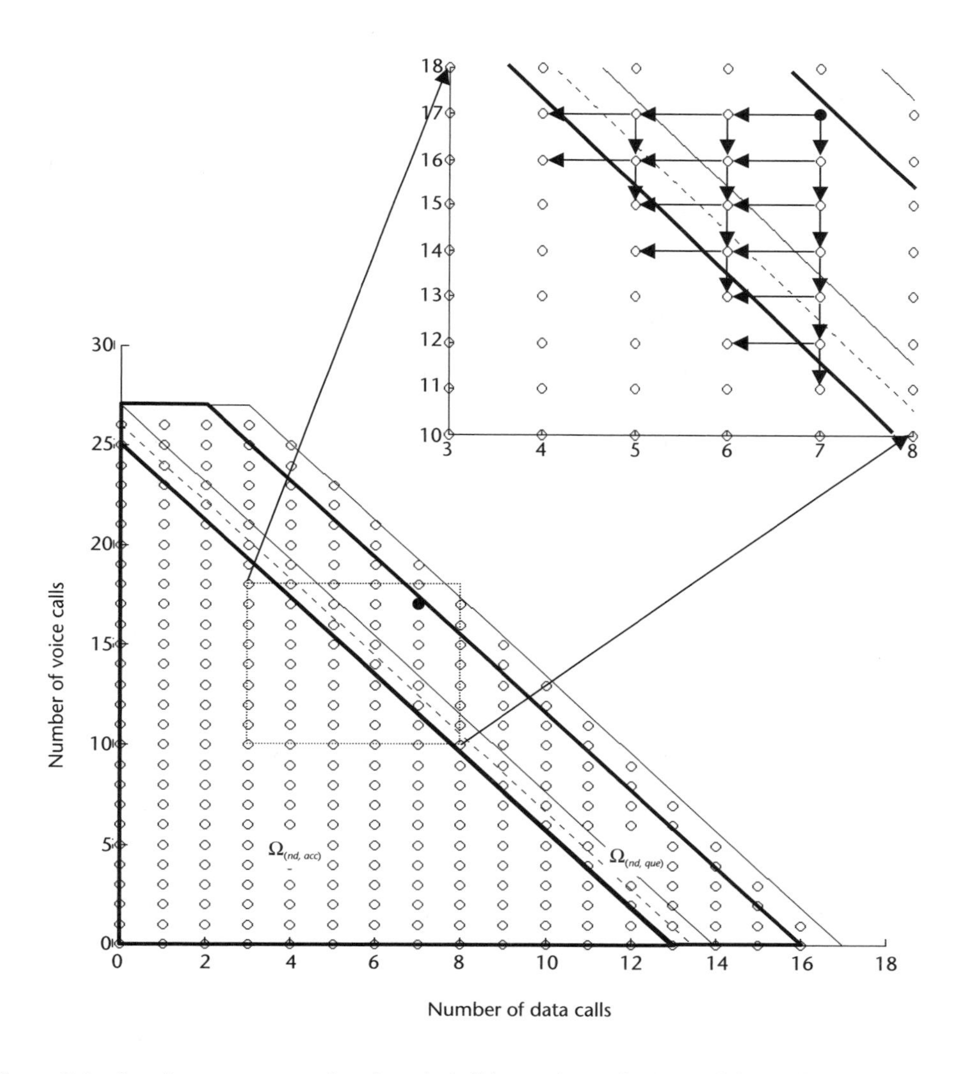

Figure 8.3 Set of states representing the admissible numbers of voice and data calls for the case that W = 1.25 MHz, q_v = 7 dB, q_d = 7 dB, R_v = 9.6 Kbps, R_d = 19.2 Kbps, and K = 3.

the case that there are 17 voice calls and 7 data calls in the system just before a new data call is admitted. In this case, the state (17, 7) can change into (i', j') in $\Omega_{(nd,acc)}$ through many paths in order for a new data call to be accepted.

For example, if (i', j') is (17,4), no voice calls and three data calls are service completed during the time τ, and if (i', j') is (16,4), one voice call and three data calls are service completed.

For the more general case where k voice calls get service completed, the delay distribution for the state (i, j) can be expressed as

$$w_{(i,j)}(\tau) = \sum_{k=0}^{I} w_{(i,j)_k}(\tau) \tag{8.25}$$

where

$$I = \min\left(i, i - \left\lfloor \frac{1-\gamma_d(1+j)}{\gamma_v} \right\rfloor\right) \tag{8.26}$$

$w_{(i,j)_k}(\tau)$ represents the delay distribution multiplied by the probability that k voice calls get service completed, given that the system state is (i, j) just before a new data call is admitted. I is the maximum number of service-completed voice calls during the change, which happens when only voice calls are service completed.

The paths where the state (i, j) in $\Omega_{(nd,que)}$ changes into (i', j') in $\Omega_{(nd,acc)}$ can be generalized as in Figure 8.4. Because the service time distribution is memoryless, and the delay distribution is independent of the current arrival, $w_{(i,j)_k}(\tau)$ is the convolution of k independent, exponential random variables, where k corresponds to the number of service-completed voice calls [9]. Because the Laplace transforms of $w_{(i,j)_k}(\tau)$ is equal to the product of the Laplace transforms of exponential distributions, the Laplace transform of $w_{(i,j)0}(\tau)$, for the case of Figure 8.4(a), can be expressed as

$$W_{(i,j)_0}(s) = \left(\frac{\left\lfloor \frac{1-\gamma_v i}{\gamma_d} \right\rfloor \mu_d}{i\mu_v + \left\lfloor \frac{1-\gamma_v i}{\gamma_d} \right\rfloor \mu_d} \right) j - \left| \frac{1-\gamma_v i}{\gamma_d} \right| + 1 \left(\frac{\left\lfloor \frac{1-\gamma_v i}{\gamma_d} \right\rfloor \mu_d}{s + \left\lfloor \frac{1-\gamma_v i}{\gamma_d} \right\rfloor \mu_d} \right)^{j - \left\lfloor \frac{1-\gamma_v i}{\gamma_d} \right\rfloor + 1} \tag{8.27}$$

The first term of $W_{(i,j)_0}(s)$ in (8.27) represents the probability for $k = 0$, which corresponds to the probability that the state (i, j) in $\Omega_{(nd,que)}$ is changed into (i', j') in $\Omega_{(nd,acc)}$ as in Figure 8.4(a). In (8.27), the exponent $\left(j - \lfloor (1-\gamma_v i)/\gamma_d \rfloor + 1\right)$ corresponds to the required number of service-completed data calls in order for the new data call to be accepted. The second term of $W_{(i,j)_0}(s)$ in (8.27) comes from the product of the Laplace transforms of exponential distributions of service time of the service-competed data calls.

For the case that $k = 1$, which corresponds to Figure 8.4(b), there are J_1 different paths and $w_{(i,j)_1}(\tau)$ is expressed as the sum of delay distributions multiplied by the probability that the path is selected out of all paths. The Laplace transform of $w_{(i,j)_1}(\tau)$ can be expressed as

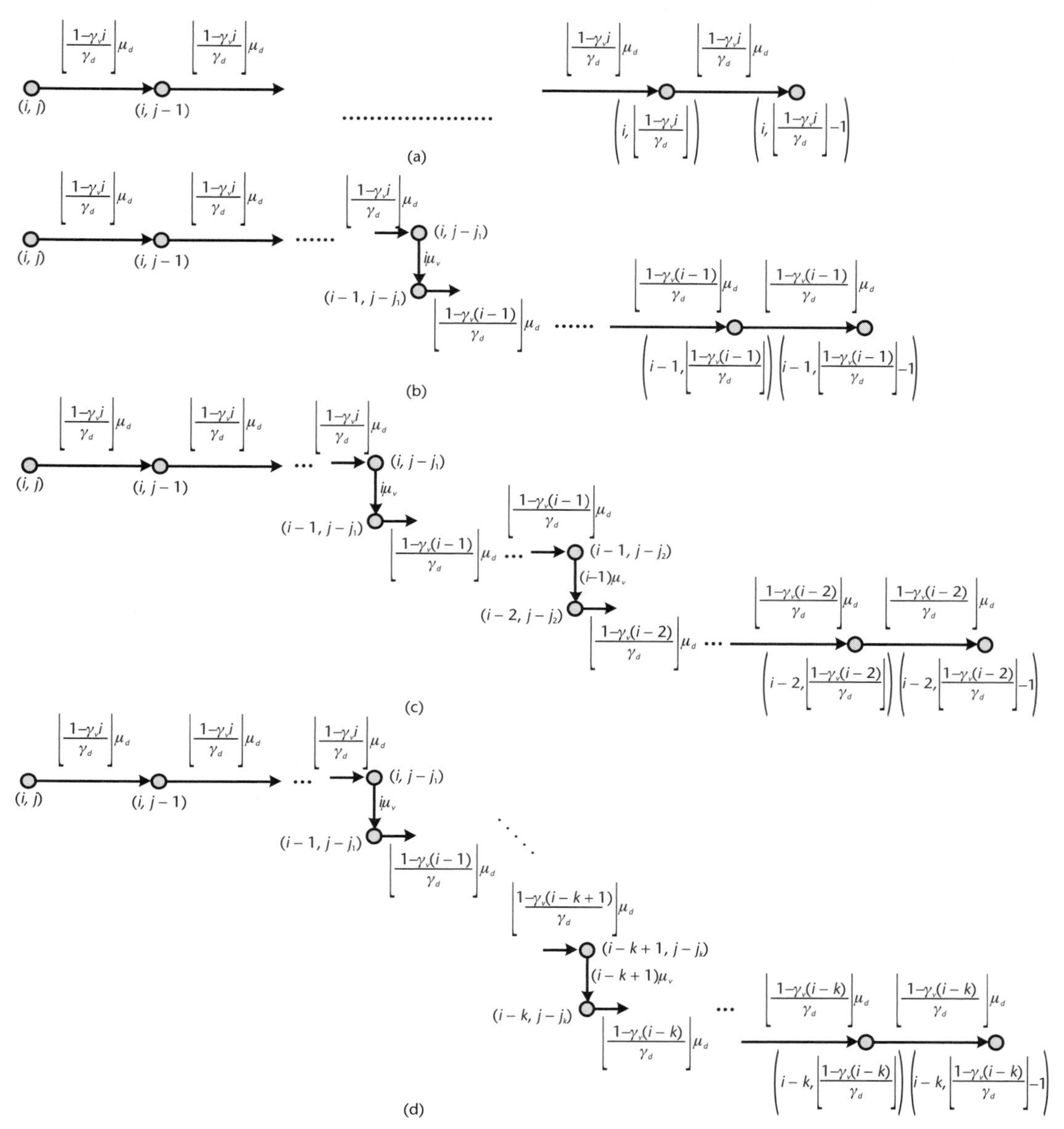

Figure 8.4 State transition paths for analyzing the delay distribution: (a) the case that no voice call and $\left(j-\lfloor(1-\gamma_v i)/\gamma_d\rfloor+1\right)$ data calls are service completed within time τ, given that the system state is (i, j) just before a new data call is attempted; (b) the case that one voice call and $\left(j-\lfloor(1-\gamma_v(i-1))/\gamma_d\rfloor+1\right)$ data calls are service completed within time ; (c) the case that two voice calls and $\left(j-\lfloor(1-\gamma_v(i-2))/\gamma_d\rfloor+1\right)$ data calls are service completed within time ; and (d) the case that k voice calls and $\left(j-\lfloor(1-\gamma_v(i-k))/\gamma_d\rfloor+1\right)$ data calls are service completed within time τ.

$$W_{(i,j)_1}(s)=\left(\frac{i\mu_v}{i\mu_v+\left\lfloor\frac{1-\gamma_v i}{\gamma_d}\right\rfloor\mu_d}\right)\left(\frac{i\mu_v}{s+i\mu_v}\right)\cdot$$

$$\sum_{j_1=0}^{J_1}\left\{\begin{array}{l}\left(\dfrac{\left\lfloor\frac{1-\gamma_v i}{\gamma_d}\right\rfloor\mu_d}{i\mu_v+\left\lfloor\frac{1-\gamma_v i}{\gamma_d}\right\rfloor\mu_d}\right)^{j_1}\left(\dfrac{\left\lfloor\frac{1-\gamma_v i}{\gamma_d}\right\rfloor\mu_d}{s_v+\left\lfloor\frac{1-\gamma_v i}{\gamma_d}\right\rfloor\mu_d}\right)^{j_1}\cdot\\ \left(\dfrac{\left\lfloor\frac{1-\gamma_v(i-1)}{\gamma_d}\right\rfloor\mu_d}{(i-1)\mu_v+\left\lfloor\frac{1-\gamma_v(i-1)}{\gamma_d}\right\rfloor\mu_d}\right)^{j-\left\lfloor\frac{1-\gamma_v(i-1)}{\gamma_d}\right\rfloor+1-j_1}\left(\dfrac{\left\lfloor\frac{1-\gamma_v(i-1)}{\gamma_d}\right\rfloor\mu_d}{s+\left\lfloor\frac{1-\gamma_v(i-1)}{\gamma_d}\right\rfloor\mu_d}\right)^{j-\left\lfloor\frac{1-\gamma_v(i-1)}{\gamma_d}\right\rfloor+1-j_1}\end{array}\right\} \quad (8.28)$$

where

$$J_1=\begin{cases} j-\left\lfloor\dfrac{1-\gamma_v(i-1)}{\gamma_d}\right\rfloor, & \text{if}\left\lfloor\dfrac{1-\gamma_v(i-1)}{\gamma_d}\right\rfloor=\left\lfloor\dfrac{1-\gamma_v i}{\gamma_d}\right\rfloor\\ j-\left\lfloor\dfrac{1-\gamma_v(i-1)}{\gamma_d}\right\rfloor+1, & \text{otherwise}\end{cases} \quad (8.29)$$

In the case of one service-completed voice call, the number of service-completed data calls should be $\left(j-\lfloor(1-\gamma_v(i-1))/\gamma_d\rfloor+1\right)$ in order for a new data call to be accepted.

J_1 is selected to avoid the path for the case of $k = 0$, and, for example, it takes the path (17, 7) → (17, 4) → (16, 4) in Figure 8.3.

By expanding the previous results to the general case of k service-completed voice calls, $W_{(i,j)_k}(s)$ can be obtained as

$$W_{(i,j)_k}(s)=$$

$$\begin{cases}\left(D_0(s)\right)^{j-\left\lfloor\frac{1-\gamma_v i}{\gamma_d}\right\rfloor}+1 & ,k=0\\ \displaystyle\prod_{a=0}^{k-1}V_a(s)\sum_{j_1=0}^{J_k}\sum_{j_2=j_1}^{J_k}\cdots\sum_{j_k=j_{k-1}}^{J_k}\left\{\begin{array}{l}\left(D_0(s)\right)^{j_1}\left(D_1(s)\right)^{j_2-j_1}\cdot\\ \left(D_2(s)\right)^{j_3-j_2}\cdots\left(D_{k-1}(s)\right)^{j_k-j_{k-1}}\cdot\\ \left(D_k(s)\right)^{j-\left\lfloor\frac{1-\gamma_v(i-k)}{\gamma_d}\right\rfloor+1-j_k}\end{array}\right\} & ,\text{otherwise}\end{cases} \quad (8.30)$$

where

$$J_k = \begin{cases} j - \left\lfloor \dfrac{1-\gamma_v(i-k)}{\gamma_d} \right\rfloor, & \text{if} \left\lfloor \dfrac{1-\gamma_v(i-k)}{\gamma_d} \right\rfloor = \left\lfloor \dfrac{1-\gamma_v(i-(k-1))}{\gamma_d} \right\rfloor, \\ j - \left\lfloor \dfrac{1-\gamma_v(i-k)}{\gamma_d} \right\rfloor + 1, & \text{otherwise} \end{cases} \quad (8.31)$$

$$V_a(s) = \left\{ \frac{(i-a)\mu_v}{(i-a)\mu_v + \left\lfloor \dfrac{1-\gamma_v(i-a)}{\gamma_d} \right\rfloor \mu_d} \right\} \left\{ \frac{(i-a)\mu_v}{s+(i-a)\mu_v} \right\} \quad (8.32)$$

and

$$D_a(s) = \left(\frac{\left\lfloor \dfrac{1-\gamma_v(i-a)}{\gamma_d} \right\rfloor \mu_d}{(i-a)\mu_v + \left\lfloor \dfrac{1-\gamma_v(i-a)}{\gamma_d} \right\rfloor \mu_d} \right) \left(\frac{\left\lfloor \dfrac{1-\gamma_v(i-a)}{\gamma_d} \right\rfloor \mu_d}{s + \left\lfloor \dfrac{1-\gamma_v(i-a)}{\gamma_d} \right\rfloor \mu_d} \right) \quad (8.33)$$

The left term of $V_a(s)$ in (8.32) is the probability that one voice call is service completed among $(i - a)$ voice and $\lfloor (1-\gamma_v(i-a))/\gamma_v \rfloor$ data calls in the service state, and the right term is a Laplace transform of the time distribution for the voice call to be service completed. On the other hand, the left term of $D_a(s)$ in (8.33) is the probability that a data call is service completed among $(i - a)$ voice and $\lfloor (1-\gamma_v(i-a))/\gamma_v \rfloor$ data calls in the service state, and the right term is a Laplace transform of the service time distribution for the data call to be service completed.

It is noteworthy that the probability of voice or data calls being service completed and the time distribution for a call to be service completed can be represented by the number of voice calls for given average service times of voice and data calls. It comes from the fact that the number of data calls in the service state is determined by the number of voice calls. $w_{(i,j)_k}(\tau)$ is the sum of delay distribution of all possible paths for k service-completed voice calls multiplied by the probability that each path is selected. J_k is a parameter to prevent $w_{(i,j)_k}(\tau)$ from including the path for $(k - 1)$ service-completed voice calls. Finally, we can get $w_{(i,j)_k}(\tau)$ from the inverse Laplace transform of $W_{(i,j)_k}(s)$.

Substituting $w_{(i,j)_k}(\tau)$ into $w_{(i,j)}(\tau)$, and then successively substituting $w_{(i,j)}(\tau)$ into $G(t)$, the CDF of delay can be calculated as

$$F_d(t) = \sum_{(i,j)\in\Omega_{(nd,acc)}} P'_{i,j} + \sum_{(i,j)\in\Omega_{(nd,que)}} \int_0^t \sum_{k=0}^{I} \mathfrak{L}^{-1}\left\{W_{(i,j)_k}(s)\right\} \cdot P'_{i,j}\, d\tau \quad (8.34)$$

where $\mathfrak{L}^{-1}$ denotes the inverse Laplace transform.

8.5 Delay Confidence

For delay-tolerant traffic, an important performance measure is related with the delay requirement. Typically, the delay requirement of data calls is given for a system to provide service with the maximum tolerable delay. Considering that the service behavior is randomly characterized, we need to introduce the delay confidence, which is defined as the probability that new data calls are accepted within the maximum tolerable delay without being blocked, and further we formulate the delay confidence as follows

$$P_c \equiv \left(1 - P_{b_d}\right) \cdot F_d\left(\tau_{\max}\right) \tag{8.35}$$

where $\tau_{\max}$ is the maximum tolerable delay. Here, note that the delay confidence is related with not only the maximum tolerable delay but also the blocking probability of data calls.

For a numerical example, we consider the system parameters in Table 8.1 and use the normalized delay, which is normalized by average service time such that

$$\tau_n = \frac{\tau}{1/\mu_d} \tag{8.36}$$

Figure 8.5 shows delay confidence as a function of the maximum tolerable delay, $\tau_{n_{\max}}$, for different offered traffic loads of voice calls when the offered traffic load of data is given as 5 and the queue size is 3. The discontinuity at $\tau_{n_{\max}} = 0$ comes from the fact that the probability that new data calls can be accepted without being blocked is nonzero. The delay confidence decreases for a fixed value of $\tau_{n_{\max}}$ as the offered traffic load of voice increases. The delay confidence increases and gradually approaches $(1 - P_{b_d})$ as the maximum tolerable delay decreases.

Figure 8.6 shows the delay confidence for different offered traffic loads of data when the offered traffic load of voice is given as 5 and the queue size is 3. The delay confidence decreases as the offered traffic load of voice increases, for a fixed value of $\tau_{n_{\max}}$. It is noteworthy that the probability that a new data call is accepted within the maximum tolerable delay without being blocked decreases as the offered traffic load

Table 8.1 System Parameters for the Numerical Example: An IS-95B-Type CDMA System Supporting Voice and Delay-Tolerable Data Services

Item	*Symbol*	*Value*
Transmission bandwidth	W	1.25 MHz
Required information data rate for voice traffic	R_v	9.6 Kbps
Required information data rate for data traffic	R_d	19.2 Kbps
Required bit energy-to-interference spectral density ratio for voice traffic	q_v	7 dB
Required bit energy-to-interference spectral density ratio for data traffic	q_d	7 dB
Average arrival rate for voice calls	λ_v	Variable
Average arrival rate for data calls	λ_d	Variable
Average service time for voice calls	$1/\mu_v$	200 seconds
Average service time for data calls	$1/\mu_d$	20 seconds

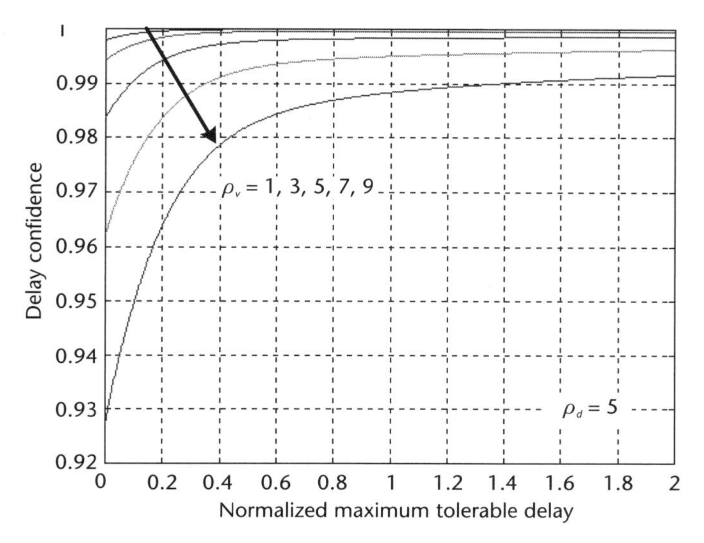

Figure 8.5 Delay confidence according to the voice traffic load when the data traffic load is given as 5.

of voice or data increases. Comparing Figure 8.6 with Figure 8.5, we can observe that the variation of delay confidence for the offered traffic load of data is greater than that for the offered traffic load of voice. This is because data calls require more system resources than do voice calls (i.e., $\gamma_d > \gamma_v$ in the case of the numerical example).

Figure 8.7 shows the delay confidence for different queue sizes when the offered traffic loads of voice and data are given as 10 and 5, respectively. The blocking probability of data calls decreases as the queue size increases. The delay confidence increases for $\tau_{n_{\max}} > 1$ while decreasing for $\tau_{n_{\max}} < 0.3$ as the queue size increases. It means that the improvement of blocking probability of data calls by means of the queue for $\tau_{n_{\max}} < 0.3$ comes from the aggravation of delay confidence. Therefore, it

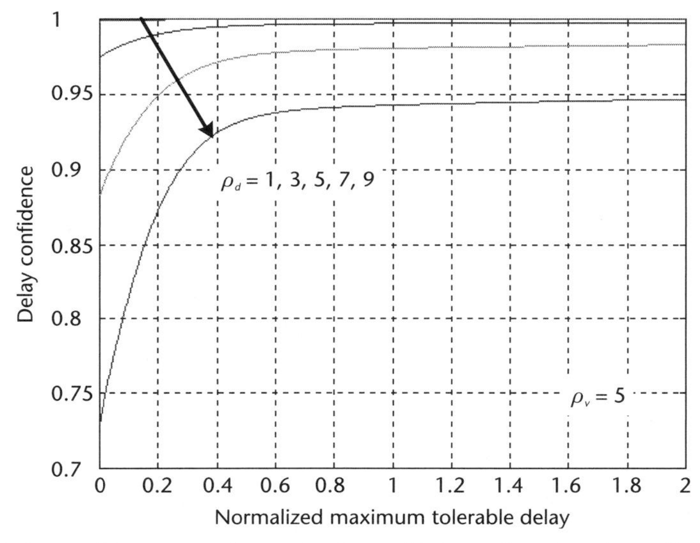

Figure 8.6 Delay confidence according to the data traffic load when the voice traffic load is given as 5.

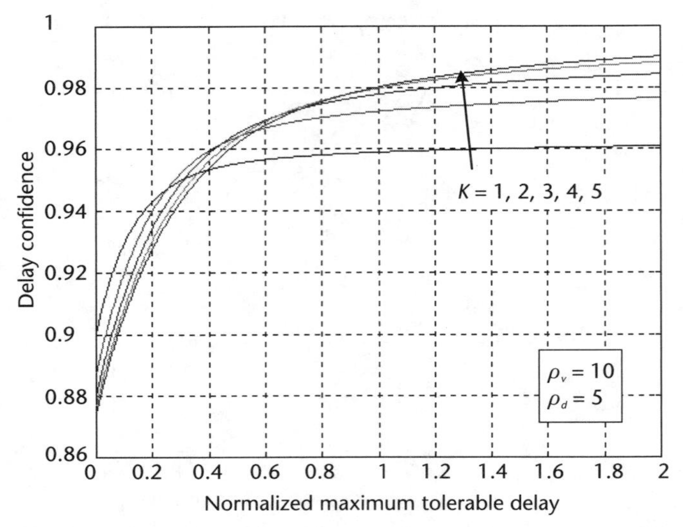

Figure 8.7 Delay confidence according to different queue sizes when the traffic load of voice and data calls is given 10 and 5, respectively.

can be expected that the queue size should be carefully selected to appropriately balance the availability of service.

8.6 Erlang Capacity

In this section, we analyze the Erlang capacity, which is defined as a set of supportable offered traffic loads of voice and data that can be supported while service requirements are satisfied. As the service requirements, we consider the required call blocking probability for voice calls and the required delay confidence for data calls. Then, the Erlang capacity of CDMA systems supporting voice and data services can be formulated as follows:

$$\begin{aligned} C_{Erlang} &\equiv \left\{(\rho_v, \rho_d) | P_{b_v} \le P_{b_{v,req}}, P_c \ge P_{c_{req}}\right\} \\ &= \left\{(\rho_v, \rho_d) | P_{b_v} \le P_{b_{v,req}}\right\} \cap \left\{(\rho_v, \rho_d) P_c \ge P_{c_{req}}\right\} \end{aligned} \tag{8.37}$$

where $P_{b_{v,req}}$ is the required blocking probability for voice calls and $P_{c_{req}}$ is the required delay confidence for data traffic. We also define a set of supportable offered traffic loads of voice and data that are confined by the required call blocking probability of voice, $P_{b_{v,req}}$, as the *voice-limited Erlang capacity* and that by the required delay confidence of data, $P_{c_{req}}$, as the *data-limited Erlang capacity*. Then, The Erlang capacity of the system is determined as the overlapped region limited by the voice-limited Erlang and the data-limited Erlang.

For certain QoS requirements and queue size, the call blocking probability of voice is a function of offered traffic loads of voice and data, and the voice-limited Erlang capacity is limited by the required blocking probability. On the other hand, the data-limited Erlang is determined by the required delay confidence and the maximum tolerable delay because the delay confidence of data traffic is a function of the maximum tolerable delay as well as the offered traffic loads of voice and data. For a numerical example, we consider the system whose parameters are given in Table 8.1.

Figure 8.8 shows the voice-limited Erlang capacity and the data-limited Erlang capacity when $K = 0$. Lines (i) and (ii) represent the voice-limited Erlang capacity and the data-limited Erlang capacity, respectively, when $P_{b_{v,req}} = 1\%$ and $P_{c_{req}} = 99\%$. The Erlang capacity is determined as the overlapped region limited by the line (i) and line (ii) to satisfy both service requirements for voice and data calls at the same time. For the case that there is no queue ($K = 0$), the CDF of delay at the maximum tolerable delay, $F_d(\tau_{max})$ is given as 1 because it is independent of the maximum tolerable delay $\tau_{n_{max}}$. In this case, the delay confidence, P_c becomes $(1 - P_{b_d})$, and the required delay confidence of 99% corresponds to the required blocking probability of 1% of data. The Erlang capacity in Figure 8.8 corresponds to that analyzed in [6] for the blocking probabilities for voice and data traffic.

Figure 8.8 also shows that the Erlang capacity is mainly determined by the data-limited Erlang capacity. The gap between the voice-limited Erlang capacity and the data-limited Erlang capacity comes from the difference in the required amount of system resources or the service requirements for voice and data calls. In this case, the data-limited Erlang capacity is lower than the voice-limited Erlang capacity for the same blocking probability because a data call requires more system resources than a voice call for this numerical example case.

In order to increase the Erlang capacity, a proper tradeoff is required between the voice-limited Erlang capacity and the data-limited Erlang capacity. One method to get the tradeoff is to use queuing for delay-tolerant data calls. Figure 8.9 shows the voice-limited Erlang capacity and the data-limited Erlang capacity for different values of the required delay confidence $P_{c_{req}}$ when $P_{b_{v,req}} = 1\%$, $\tau_{n_{max}} = 0.1$, and $K = 1$.

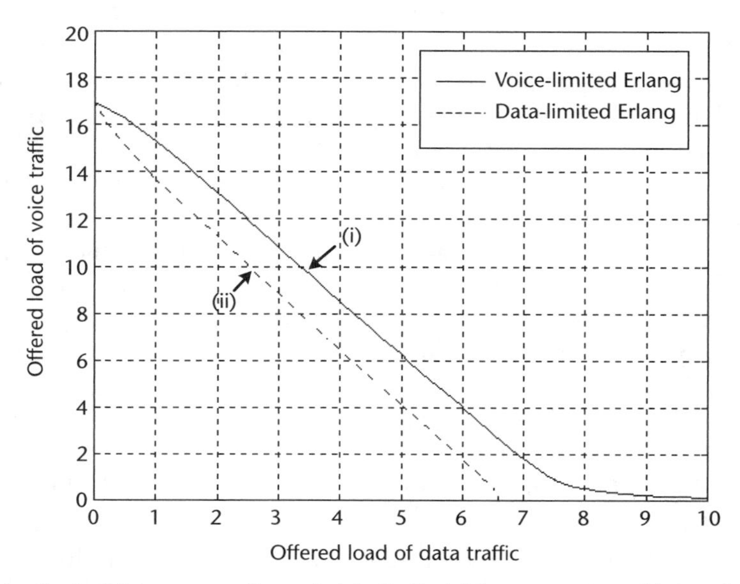

Figure 8.8 Voice-limited Erlang capacity and data-limited Erlang capacity without allowing the delay in queue; (i) and (ii) represent the voice-limited Erlang and the data-limited Erlang, respectively, and the Erlang capacity corresponds to the overlapped region limited by Erlang capacity lines (i) and (ii) where $P_{b_{v,req}} = 1\%$, $P_{c_{req}} = 99\%$, and $K = 0$.

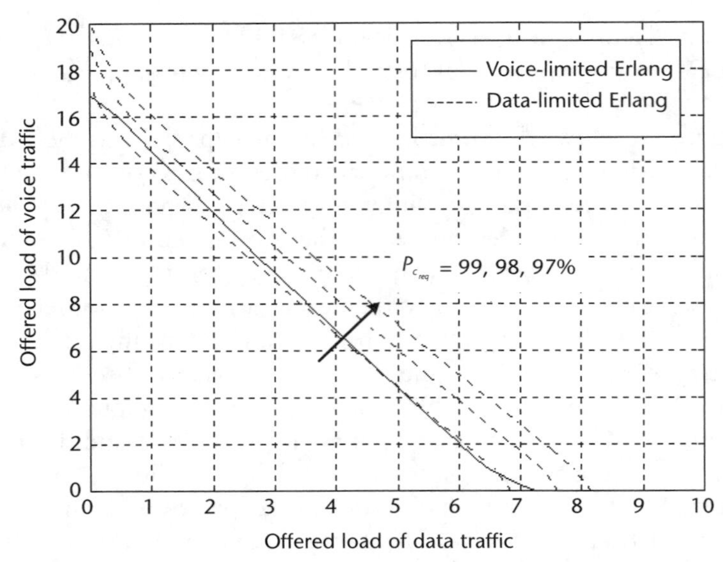

Figure 8.9 Voice-limited Erlang capacity and data-limited Erlang capacity for different values of the required delay confidence $P_{c_{req}}$ when $P_{b_{v,req}} = 1\%$, $\tau_{n_{max}} = 0.1$, and $K = 1$.

In this case, a new data call can be queued until the required resources are available if the queue is available. In this case, the Erlang capacity is mainly determined by the data-limited Erlang capacity when $P_{c_{req}} = 99\%$. The data-limited Erlang capacity gradually increases as the required delay confidence $P_{c_{req}}$ decreases such that the Erlang capacity is determined by the voice-limited Erlang capacity when $P_{c_{req}}$ is given less than 98%.

Figure 8.10 shows the voice-limited Erlang capacity and the data-limited Erlang capacity for different values of the maximum tolerable delay $\tau_{n_{max}}$ when $P_{b_{v,req}} = 1\%$, $P_{c_{req}} = 99\%$, and $K = 1$. The Erlang capacity is mainly determined by the data-limited Erlang capacity when the maximum tolerable delay $\tau_{n_{max}}$ is less than 0.1. As $\tau_{n_{max}}$ increases (i.e., when the delay allowance for data calls increases), the data-limited Erlang capacity also increases. Figure 8.10 shows that the Erlang capacity is determined by the voice-limited Erlang capacity when $\tau_{n_{max}}$ is more than 0.4.

From Figures 8.9 and 8.10, we know that the delay requirements such as the required delay confidence and the maximum tolerable delay have no effect on the Erlang capacity beyond certain limits. It comes from the fact that the voice-limited Erlang capacity is independent of the delay confidence and the maximum tolerable delay. As the delay requirements become looser, the data-limited Erlang capacity increases while the voice-limited Erlang capacity does not change. The gain from the delay requirements in the data-limited Erlang capacity over the voice-limited Erlang capacity cannot be supported by the system because the required call blocking probability of voice is not guaranteed.

Figure 8.11 shows the effect of the queue size on the Erlang capacity when $P_{b_{v,req}} = 1\%$, $P_{c_{req}} = 99\%$, and $\tau_{n_{max}} = 0.1$. The solid and dotted lines represent the voice-limited Erlang capacity and data-limited Erlang capacity, respectively. We know that the voice-limited Erlang capacity decreases as the queue size increases, which comes from the fact that the call blocking probability of voice increases for a larger queue size. On the other hand, the data-limited Erlang capacity also increases until

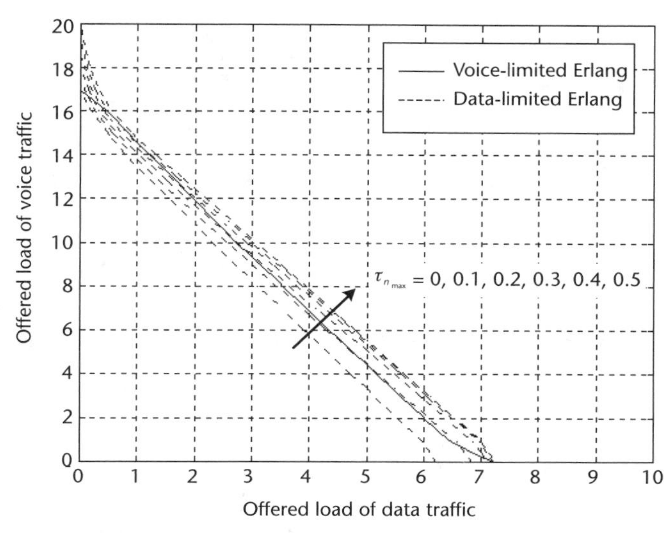

Figure 8.10 Voice-limited Erlang capacity and data-limited Erlang capacity for different values of the maximum tolerable delay $\tau_{n_{max}}$ when $P_{b_{v,req}}$ = 1%, $P_{c_{req}}$ = 99%, and $K = 1$.

the queue size becomes 2. After that, it decreases for a larger queue size. This comes from the fact that the call blocking probability of data and the CDF of delay decrease as the queue size increases. Noting that the delay confidence is enhanced as the blocking probability of data calls decreases or the CDF of delay increases, we know that the change from an increase to a decrease of data-Erlang capacity according to the queue size results from the mutual effects between the improvement in the blocking probability of data calls and the decrease of the CDF of delay, which also can be observed in Figure 8.7. Figure 8.11 also shows that the Erlang capacity when $K = 3$ is less than that when $K = 0$. This means that the queue size should be properly selected to create a balance between the voice-limited Erlang capacity and the data-limited Erlang capacity and further to accommodate more Erlang capacity. In the numerical example, the optimum queue size can be selected as 1, with respect to the Erlang capacity.

8.7 Conclusions

In this chapter, we analyzed the Erlang capacity of a CDMA system supporting voice and delay-tolerant data services, and considered the characteristic of delay-tolerant traffic, called the delay confidence, which is defined as the probability that a new data call is accepted within the maximum tolerable delay without being blocked. For the performance analysis, we developed a two-dimensional Markov chain model, based on the FCFS service discipline, and presented a numerical procedure to analyze the Erlang capacity. As a result, for the case that there is no queue for data calls, it was observed that the Erlang capacity is mainly determined by the data-limited Erlang capacity, as one data call requires more system resources than one voice call. For the case that we consider finite-size buffer for data calls, the

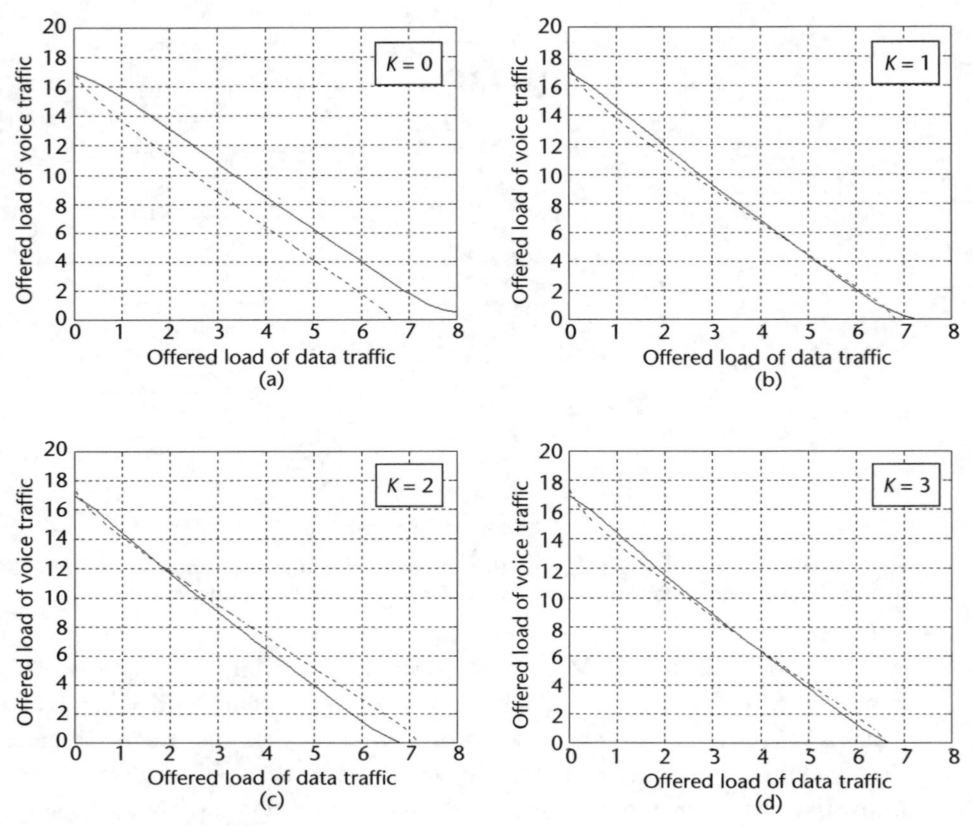

Figure 8.11 Effect of the queue size on the Erlang capacity when $P_{b_{v,req}} = 1\%$, $P_{c_{req}} = 99\%$, and $\tau_{n_{\max}} = 0.1$: (a) $K = 0$, (b) $K = 1$, (c) $K = 2$, and (d) $K = 3$.

data-limited Erlang capacity increases as the maximum tolerable delay increases or the required delay confidence decreases. Further, the Erlang capacity is mainly limited by the voice-limited Erlang capacity if the required delay confidence and the maximum tolerable delay requirements go beyond certain limits.

By observing the Erlang capacity according to the queue size, we showed that the queue size should be properly selected to create a balance between the voice-limited Erlang capacity and the data-limited Erlang capacity. For the numerical example case, we demonstrated that a proper queue size was selectable with respect to the Erlang capacity under a given delay constraint.

References

[1] Viterbi, A. M., and A. J. Viterbi, "Erlang Capacity of a Power-Controlled CDMA System," *IEEE Journal on Selected Areas in Communications*, 1993, pp. 892–900.

[2] Sampath, A., N. B. Mandayam, and J. M. Holtzman, "Erlang Capacity of a Power Controlled Integrated Voice and Data CDMA System," *IEEE Proc. of Vehicular Technology Conference*, 1997, pp. 1557–1561.

[3] Sampath, A., P. S. Kumar, and J. M. Holtzman, "Power Control and Resource Management for a Multimedia CDMA Wireless System," *IEEE Proc. of International Symposium on Personal, Indoor and Mobile Radio Communications*, 1995, pp. 21–25.

[4] Yang, Y. R., et al., "Capacity Plane of CDMA System for Multimedia Traffic," *IEEE Electronics Letters*, 1997, pp. 1432–1433.

[5] Koo, I., E. Kim, and K. Kim, "Erlang Capacity of Voice/Data DS-CDMA Systems with Prioritized Services," *IEICE Trans. on Communications*, 2001, pp. 716–726.

[6] Cruz-Perez, F. A., and M. L. D. Lara-Rodriguez, Performance Analysis of the Fractional Channel Reservation in TDMA and CDMA Integrated Services Networks," *IEEE Proc. of Vehicular Technology Conference*, Spring 2001, pp. 1007–1011.

[7] Bae, B. S., K. T. Jin, and D. H. Cho, "Performance Analysis of an Integrated Voice/Data CDMA System with Dynamic Admission/Access Control," *IEEE Proc. of Vehicular Technology Conference*, Spring 2001, pp. 2440–2444.

[8] Ishikawa, Y., and N. Umeda, "Capacity Design and Performance of Call Admission Control in Cellular CDMA Systems," *IEEE Journal on Selected Areas in Communications*, 1997, pp. 1627–1635.

[9] Gross, D., and C. M. Harris, *Fundamentals of Queueing Theory,* New York: John Wiley & Sons, 1998.

CHAPTER 9

Multiclass CDMA Systems with a Limited Number of Channel Elements

Because the CDMA system is the interference-limited system, directional antennas are usually used for spatial isolation, which reduces interference. In a multisectorized cell using directional antennas, the call blocking occurs not only due to the insufficient number of CEs available for traffic channels in BSs but also due to a limit on the number of concurrent users in each sector [1]. The CE performs the baseband spread-spectrum signal processing for a given channel. For trunking efficiency, all CEs are provided per cell, and any CE can be assigned to any user in the cell regardless of sector. Call blocking due to insufficient CEs available for traffic channels in the BS is defined as *hard blocking*. So far, we assumed that the CDMA system of our interest has a sufficient number of CEs and the system only suffers from *soft* capacity issues. For example, in Chapters 7 and 8, the Erlang capacity of CDMA systems is already investigated based on only *soft blocking* for the case of multiclass services and the voice and data services under the delay constraint, respectively.

In this chapter, we will investigate the effect of a limited number of CEs in BSs on the Erlang capacity of CDMA systems supporting multiclass services as an expansion work of Chapters 7 and 8. In addition, a graphic interpretation method also will be presented for the case of multiple FAs, where the required calculation complexity of the exact method is too high to calculate the Erlang capacity. In next chapter, we will address an approximation method to calculate the Erlang capacity of CDMA systems with a limited number of CEs in BSs to overcome the complexity problem of the exact calculation method presented in this chapter.

9.1 Introduction

In CDMA systems, unlike FDMA or TDMA systems, a call attempt may be blocked due not only to the insufficient number of CEs available for traffic channels but also to the excess of the maximum allowable number of concurrent users. In a CDMA system, the CE performs the baseband spread spectrum signal processing for a given channel. At a sectorized cell, all CEs are shared in the BS for the trucking efficiency such that any CE can be assigned to any user in the cell, regardless of sector. Call blocking, which is caused by insufficient CEs in the BS, is hard blocking. Additionally, in the CDMA, excessive interference also causes a call blocking. This is soft blocking and occurs when the number of active users exceeds the maximum allowable number of concurrent users. Research to find the maximum allowable number of concurrent users that CDMA can support in the reverse link has been done in [2–4], based on the maximum tolerable interference.

For the purpose of controlling the system, another measure of system capacity is the peak average load that can be supported with a given quality and with availability of service as measured by the blocking probability. The average traffic load in terms of the average number of users requesting service resulting in the required blocking probability is called the *Erlang capacity*. In [5], Viterbi and Viterbi reported that the Erlang capacity of a CDMA system supporting only voice service, based on outage probability, where the outage probability is defined as the probability that the interference plus noise power density, exceeds the noise power density N_o by a factor of $1/\eta$, where η takes on typical values between 0.25 and 0.1. In [6], Sampath et al. extended the results of Viterbi to voice/data CDMA systems.

Furthermore, Matragi et al. introduced another approach that allows the provision of different GoS for different types of calls [7]. It is noteworthy that the aforementioned analysis considered the soft blocking only, and the effect of CEs on Erlang capacity was not considered. Practically, the CDMA system is equipped with a finite number of CEs, afforded by its cost-efficient strategy, which introduces inherent hard blocking. Between soft blocking and hard blocking, the former has been analyzed completely in [8] or Chapter 7 for the multimedia CDMA, while it is an interesting question to the system operator to evaluate the exact effect of the Erlang capacity due to the limited number of CEs (i.e., hard blocking).

Subsequently, in this chapter, we present an analytical procedure for deriving the Erlang capacity of CDMA systems supporting multimedia services in the reverse link, by considering hard blocking as well as soft blocking when the CDMA cells are sectorized with three sectors.

The remainder of this chapter is organized as follows. In Section 9.2, we describe the system model and briefly summarize the capacity bound of the maximum allowable number of concurrent users that CDMA systems can support with QoS requirements. In Section 9.3, an analytical procedure for analyzing the Erlang capacity of the multimedia CDMA systems is presented, based on the multidimensional Markov model. In Section 9.4, a numerical example is taken into consideration, and discussions are given. Finally, conclusions are drawn in Section 9.5.

9.2 System Model

The system we are considering employs a circuit switching method to deal with the data transmission, as handled in [8]. Furthermore, we assume that the perfect directional antennas are used whereby the cell is partitioned into three 120° sectors, where all available CEs in the BS are shared among three sectors such that any CE can be assigned to any user in the cell, regardless of sector.

As a reference for soft blocking at each sector, in this chapter, a capacity limit of the maximum allowable number of concurrent users that CDMA can support with QoS requirements in the reverse link is utilized. In the case of CDMA, although there is no hard limit on the number of mobile users served, there is a practical limit on the number of concurrent users to control the interference between users having the same pilot signal. The maximum allowable number of concurrent users that a CDMA system can support with QoS requirements has been found [2–4], based on the maximum tolerable interference. This issue has already been dealt with in Chapter 2. In particular, as a result of [4], the system capacity limit of CDMA system supporting K district traffic types (one voice and $K - 1$ data service groups) in the reverse link can be expressed as:

$$\gamma_v n_v + \sum_{j=1}^{K-1} \gamma_{d_j} n_{d_j} \leq 1 \tag{9.1}$$

where

$$\gamma_v = \frac{\alpha}{\frac{W}{R_{v_{req}}} \left(\frac{E_b}{N_o}\right)_{v_{req}}^{-1} \left\langle \frac{1}{1+f} \right\rangle 10^{\frac{Q^{-1}(\beta)}{10}\sigma_x - 0.012\sigma_x^2} + \alpha}$$

$$\gamma_{d_j} = \frac{1}{\frac{W}{R_{d_j,req}} \left(\frac{E_b}{N_o}\right)_{d_j,req}^{-1} \left\langle \frac{1}{1+f} \right\rangle 10^{\frac{Q^{-1}(\beta)}{10}\sigma_x - 0.012\sigma_x^2} + 1}$$

All relevant parameters are defined and described in Section 3.1.

The inequality of (9.1) is the necessary and sufficient condition satisfying the system QoS requirements and indicates that calls of different services take different amount of system resources according to their QoS requirements (e.g., information data rate and the required bit energy-to-inference power spectral density ratio). In the following analysis, based on (9.1), we assume that one call attempt of data in the jth service group is equivalent to Λ_j call attempts of voice service, where Λ_j is defined as $\lfloor \gamma_{d_j} / \gamma_v \rfloor$ where $\lfloor x \rfloor$ denotes the greatest integer is less than or equal to x. Then, (9.1) can be rewritten as follows:

$$n_v + \sum_{j=d_1}^{d_{K-1}} \Lambda_j \cdot n_j \leq \hat{C}_{ETC} \tag{9.2}$$

where $\hat{C}_{ETC} \equiv \lfloor 1 / \gamma_v \rfloor$ is the total number of basic channels, and subscript "ETC" denotes equivalent telephone (voice) channel. That is, the voice channel is presumed to the basic channel. The system capacity limit, stipulated by (9.1) or (9.2), can be considered as the possible number of concurrent calls per sector that can be managed on the reverse link while the QoS requirements being satisfied.

9.3 Erlang Capacity for the Multimedia CDMA Systems

In this section, we will analyze the Erlang capacity, based on the multidimension *M/M/m* loss model. To analyze the Erlang capacity, the state probability of the system will be developed and the call blocking probabilities experienced by each call will be found by summing the occupation probabilities of the corresponding call blocking states.

For the performance analysis, we assume that call attempts of the K district traffic types at each sector are generated according to mutually independent Poisson processes with rates $\lambda_{(j,i)}$ and require Λ_j basic channels. In addition, they have the channel holding times, which are exponentially distributed with mean $1/\mu_{(j,i)}$, where the subscription of i denotes the ith sector and the subscription of j indicates the jth service group ($j = v, d_1, \ldots, d_{K-1}$; $i = 1, 2, 3$).

Let $\mathbf{N}_i = (n_{(v,i)}, n_{(d_1,i)}, \ldots, n_{(d_{K-1},i)})$ be the state of the ith sector ($i = 1, 2, 3$), given by the number of calls of each service group in the ith sector, and assume that a capacity bound stipulated in (9.1) is used as a reference to threshold for soft blocking. Then, the state probability of $\mathbf{N}_i$ in the ith sector is given by [9]:

$$\pi_i(\mathbf{N}_i) = \frac{1}{G_i(R)} \prod_{j=v}^{d_{K-1}} \frac{\rho_{(j,i)}^{n_{(j,i)}}}{n_{(j,i)}!} \quad \text{for } \mathbf{N}_i \in \Omega_i(R) \text{ and } i = 1, 2, 3 \tag{9.3}$$

where $\rho_{(j,i)} = \lambda_{(j,i)}/\mu_{(j,i)}$, which denotes the offered traffic load of the jth service group in the ith sector. $G_i(R)$ is a normalizing constant that has to be calculated in order to have the $\pi_i(\mathbf{N}_i)$ that is accumulated to 1:

$$G_i(R) = \sum_{\mathbf{N}_i \in \Omega_i(R)} \prod_{j=v}^{d_{K-1}} \frac{\rho_{(j,i)}^{n_{(j,i)}}}{n_{(j,i)}!} \tag{9.4}$$

For a multimedia CDMA system supporting K service groups, as we described in the previous section, a set of admissible states in the ith sector, $\Omega_i(R)$, can be given as

$$\Omega_i(R) = \{\mathbf{N}_i \mid \mathbf{N}_i \mathbf{A}^T \le R\} \tag{9.5}$$

where $\mathbf{N}_i$ and $\mathbf{A}$ are 1 by K vector and R is a scalar representing the system resource such that

$$\mathbf{A} = (1, \Lambda_{d_1}, \ldots, \Lambda_{d_{K-1}}) \text{ and } R = \hat{C}_{ETC} \tag{9.6}$$

To analyze the cell as a whole, it is useful to define occupation state of the cell S characterized by the occupation numbers of the sectors as a state in the birth-death process. That is,

$$S \in \{(\mathbf{N}_1, \mathbf{N}_2, \mathbf{N}_3) \mid \mathbf{N}_1 \in \Omega_1(R), \mathbf{N}_2 \in \Omega_2(R) \text{ and } \mathbf{N}_3 \in \Omega_3(R)\} \tag{9.7}$$

Because traffic for the individual sectors can be assumed to be independent processes, the state probability $\pi(S)$ that the multidimensional Markov chain is in the state of S is the product of the individual sector probabilities, such that

$$\pi(S) = \pi_1(\mathbf{N}_1) \cdot \pi_2(\mathbf{N}_2) \cdot \pi_3(\mathbf{N}_3) \tag{9.8}$$

As mentioned before, at the BS, all of the CEs are shared, so any CE can be assigned to any user in the cell, regardless of sector. Also, hard blocking occurs when the number of CEs that are used by concurrent users exceeds the maximum number of available CEs in the BS. Such effect of the limitation of CEs in the BS on the call blocking can be considered by adding the constraint of CEs to (9.7), such that

$$S' \in \left\{ \begin{array}{l} (\mathbf{N}_1, \mathbf{N}_2, \mathbf{N}_3) | \mathbf{N}_1 \in \Omega_1(R), \mathbf{N}_2 \in \Omega_2(R), \mathbf{N}_3 \in \Omega_3(R) \\ \text{and} \sum_{i=1}^{3} n_{(v,i)} + \sum_{j=d_1}^{d_{K-1}} \sum_{i=1}^{3} n_{(j,i)} \cdot \Lambda_j \le N \end{array} \right\} \tag{9.9}$$

where N is the maximum number of available CEs in the BS. Here, it is assumed that a number of CEs used by one user in the jth service group is directly proportional to Λ_j, even though it depends on the modem structure of the system being considered.

Because the constraint of (9.9) limits the total number of users of each service group, the state probability $\pi(S')$ can be derived from the joint conditional density function of $\mathbf{N}_1$, $\mathbf{N}_2$, and $\mathbf{N}_3$, given $\left\{ \sum_{i=1}^{3} n_{(v,i)} + \sum_{j=d_1}^{d_{K-1}} \sum_{i=1}^{3} n_{(j,i)} \cdot \Lambda_j \le N \right\}$.

Namely,

$$\pi(S') = \frac{\pi_1(\mathbf{N}_1) \cdot \pi_2(\mathbf{N}_2) \cdot \pi_3(\mathbf{N}_3)}{C} \tag{9.10}$$

where

$$\begin{aligned} C &\equiv P\left(\sum_{i=1}^{3} n_{(v,i)} + \sum_{j=d_1}^{d_{K-1}} \sum_{i=1}^{3} n_{(j,i)} \cdot \Lambda_j \le N \right) \\ &= \sum_{\mathbf{N}_1 \in \Omega_1(R)} \sum_{\mathbf{N}_2 \in \Omega_2(R)} \frac{G_3\left(N - \sum_{i=1}^{2} \mathbf{N}_i \mathbf{A}^T\right)}{G_3(R)} \cdot \pi_1(\mathbf{N}_1) \cdot \pi_2(\mathbf{N}_2) \end{aligned} \tag{9.11}$$

Note that C corresponds to the probability of the event $\left\{ \sum_{i=1}^{3} n_{(v,i)} + \sum_{j=d_1}^{d_{K-1}} \sum_{i=1}^{3} n_{(j,i)} \cdot \Lambda_j \le N \right\}$, which assures that the probabilities of the valid states sum to 1. Also, note that $C = 1$ if $N > 3\hat{C}_{ETC}$.

Generally, the call blocking probabilities of each service group in each sector can be found by summing the occupation probabilities of the corresponding call blocking states. The call blocking states for each service group in each sector are mainly separated into two parts: soft-blocking and hard-blocking states. For the jth service group ($j = v, d_1, \ldots, d_{K-1}$) at the first sector, the call blocking states are given as follows:

$$\Omega_{(b,soft)} = \left\{ \begin{array}{l} S' | \hat{C}_{ETC} - \Lambda_j < \mathbf{N}_1 \mathbf{A}^T \le \hat{C}_{ETC} \\ \mathbf{N}_2 \in \Omega_2(R), \mathbf{N}_3 \in \Omega_3(R) \text{ and} \sum_{i=1}^{3} \mathbf{N}_i \mathbf{A}^T < N \end{array} \right\} \tag{9.12}$$

$$\Omega_{(b,hard)} = \left\{ \begin{array}{l} S' | N - \Lambda_j - 1 \le \sum_{i=1}^{3} \mathbf{N}_i \mathbf{A}^T \le N, \\ \mathbf{N}_1 \in \Omega_1(R), \mathbf{N}_2 \in \Omega_2(R) \text{ and } \mathbf{N}_3 \in \Omega_3(R) \end{array} \right\} \quad (9.13)$$
$$for\ j = v, d_1, \ldots, d_{K-1}$$

Then, the corresponding call blocking probability for the jth service group is given as

$$\begin{aligned} P_{(blocking,j)} &= P_{b(soft,j)} + P_{b(hard,j)} \\ &= \sum_{S' \in \Omega_{(b,soft)}} \pi(S') + \sum_{S' \in \Omega_{(b,hard)}} \pi(S') \end{aligned} \quad (9.14)$$

Here, note that if $N > 3\hat{C}_{ETC}$, $P_{b(soft,j)}$, is simplified as

$$P_{b(soft,j)} = 1 - \frac{G_1(R - \mathbf{A}e_j)}{G_1(R)} \quad (9.15)$$

where e_j is a unit vector in the jth direction and $G_1(R)$ is the normalizing constant calculated on the whole $\Omega_1(R)$, while $G_1(R - \mathbf{A}e_j)$ is the normalizing constant calculated on the $\Omega_1(R - \mathbf{A}e_j)$ with respect to the traffic of the jth service group.

Also, $P_{b(j,hard)} = 0$ if $N \ge 3\hat{C}_{ETC} + \Lambda_j$ for the traffic of the jth service group. Additionally, for all service groups, soft-blocking states do not exist if $N \le \hat{C}_{ETC}$, and the call blocking is determined only by hard blocking.

Here note that, even though we have only presented the procedures for evaluating the call blocking probability of the jth service group at the first sector, similar analysis can be performed for calculating the call blocking probabilities of each service group at the second and third sectors.

In the multimedia environment, Erlang capacity corresponding to the voice-only system needs to be modified in order to consider the performance of all service groups simultaneously. In this chapter, a modified Erlang capacity is utilized as a performance measure. It is defined as a set of the average offered traffic loads of each service group that can be supported while the QoS and GoS requirements being satisfied. Then, Erlang capacity per sector can be calculated as follows:

$$\begin{aligned} &C_{Erlang} \\ &= \left\{ \left(\hat{\rho}_v, \hat{\rho}_{d_1}, \ldots, \hat{\rho}_{d_{K-1}} \right) \right\} \\ &= \left\{ \begin{array}{l} \left(\rho_v, \rho_{d_1}, \ldots, \rho_{d_{K-1}} \right) | P_{(blocking,v)} \le P_{(B,v)_{req}}, P_{(blocking,d_1)} \le P_{(B,d_1)_{req}}, \\ \ldots, P_{(blocking,d_{K-1})} \le P_{(B,d_{K-1})_{req}} \end{array} \right\} \end{aligned} \quad (9.16)$$

where $P_{(B,v)_{req}}$, $P_{(B,d_1)_{req}}$, ..., $P_{(B,d_{K-1})_{req}}$ are the required call blocking probabilities of voice and $K-1$ data service groups, respectively, which can be considered requirements.

In other words, the system Erlang capacity is the set of values of $\{(\hat{\rho}_v, \hat{\rho}_{d_1}, \ldots, \hat{\rho}_{d_{K-1}})\}$ that keeps the call blocking probability experienced by each call less than the required call blocking probability (or GoS requirements). Under these conditions, the Erlang capacity with respect to the jth service group can be calculated as a function of the offered traffic loads of all service groups by contouring the call blocking probability experienced by the traffic of the jth service group at the required call blocking probability. Finally, total system Erlang capacity is determined by the overlapped region of Erlang capacities with respect to all service groups. An easy way to visualize total system Erlang capacity is to consider the overlapped Erlang capacity region as total system Erlang capacity. Consequently, it is necessary to balance the Erlang capacities with respect to all service groups and to get the proper tradeoff in order to enhance total system Erlang capacity.

9.4 Numerical Example and Discussion

9.4.1 Single FA Case

First, let's consider a typical IS-95B CDMA system supporting voice and data traffic with single FA whose frequency bandwidth is 1.25 MHz. For three-sector CDMA cells, assuming that the sectors are equally loaded, the system parameters under the consideration are shown in Table 9.1. The data rate is 28.8 Kbps for data traffic by aggregating three codes, and 9.6 Kbps for voice. Also, 100% activity factor is assumed for data. In this case, Λ and $\hat{C}_{ETC}$ are given as 6 and 29, respectively, based on (9.1). It means that there are 29 basic channels per sector, and one call attempt of data traffic is equivalent to six call attempts of voice traffic. This section provides calculated the Erlang capacity per sector.

For the numerical example, the Erlang capacity per sector is depicted in two-dimensional space and is given as the set of the offered traffic loads of voice and data in which the call blocking probabilities of voice and data are maintained below the required call blocking probabilities of voice and data. Figure 9.1 shows the Erlang

Table 9.1 System Parameters for the IS-95B-Type CDMA System Supporting Voice and Data Services

Parameters	*Symbol*	*Value*
Allocated frequency bandwidth	W	1.25 Mbps
Required bit transmission rate for voice traffic	R_v	9.6 Kbps
Required bit transmission rate for data traffic	R_d	28.8 Kbps
Required bit energy-to-interference power spectral density ratio for voice traffic	$\left(\frac{E_b}{N_o}\right)_{v\,req}$	7 dB
Required bit energy-to-interference power spectral density ratio for data traffic	$\left(\frac{E_b}{N_o}\right)_{d\,req}$	7 dB
System reliability requirement	$\beta\%$	99%
Frequency reuse factor	$\left\langle\frac{1}{1+f}\right\rangle$	0.7
Standard deviation of received SIR	σ_x	1 dB
Voice activity factor	α	3/8

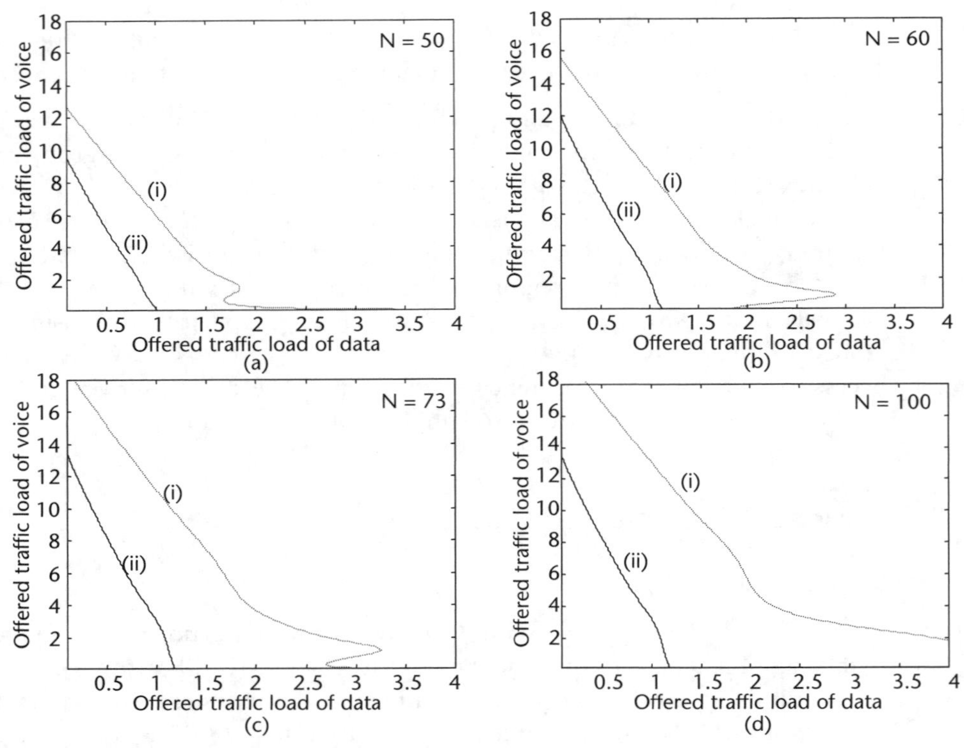

Figure 9.1 Erlang capacity for the different values of CEs when the required call blocking probabilities (GoS) of voice and data are given as (2%), respectively: (a) when $N = 50$, (b) when $N = 60$, (c) when $N = 73$, (d) when $N = 100$. For each case, the curve represented by (i) is the Erlang capacity with respect to voice traffic, and the curve represented by (ii) is the Erlang capacity with respect to data traffic.

capacity per sector for different values of CEs when the required call blocking probabilities (or GoS requirements) for voice and data traffic are given as 2%. For each case, the curve represented by (i) is the Erlang capacity with respect to voice traffic, and the curve represented by (ii) is the Erlang capacity with respect to data traffic. When $N = 50$, 60, or 73, it is noteworthy that there is the "winding characteristic," in which the Erlang capacity curves with respect to voice traffic become winded in the region of low voice traffic load.

This phenomenon results from the fact that the blocking probability experienced by the voice traffic is suddenly degraded for certain traffic loads because the data traffic is blocked unless a set of Λ basic channels is available, while the voice traffic can be served if a basic channel is available. Additionally, this phenomenon is generally observed when the load of voice traffic is very low and the blocking probability experienced by the voice traffic is very susceptible to the change of the data traffic load. In addition, Figure 9.1 indicates that the more CEs there are, the larger the Erlang capacity is.

For a fixed number of CEs, the following observations are made. The first is that data traffic has more of an impact than voice traffic on Erlang capacity because the effective bandwidth required by one data user is larger than that of one voice user. The other observation is that the total system Erlang capacity region is determined by the Erlang capacity with respect to data traffic because the system should satisfy the required call blocking probabilities of voice and data groups simultaneously. Hence, it is required to get the proper tradeoff between Erlang capacities with respect to voice and data traffic in order to enhance total system Erlang capacity. One way to consider this tradeoff is to give data traffic priority over voice traffic by using prioritized schemes, such as a reservation scheme in which some system resources can be exclusively reserved for the data traffic. Another way is to provide different GoS requirements for voice and data traffic. In this case, call blocking probability higher than 2% can be given for data traffics.

Figure 9.2 shows the Erlang capacity for the different GoS requirements for voice and data traffic. For each case, the curve represented by (i) is the Erlang capacity with respect to voice traffic, the curve represented by (ii) is the Erlang capacity with respect to data traffic, and the curve represented by (iii) is the Erlang capacity with respect to average GoS. For any traffic load of voice and data, the call blocking probability experienced by the data traffic is always higher than that of voice traffic. Hence, as mentioned before, the total system Erlang capacity region is determined by the Erlang capacity with respect to data traffic when same GoS requirements are given for voice and data traffic. To solve such problems, careful selection of operating values of GoS requirements for voice and data traffic is needed. Here, three cases for selecting the proper operating values of GoS requirements are given. First, we consider the case of a 2% GoS requirement for both voice and data traffic. As a merit of this case, the strict GoS requirements can be satisfied for both types of

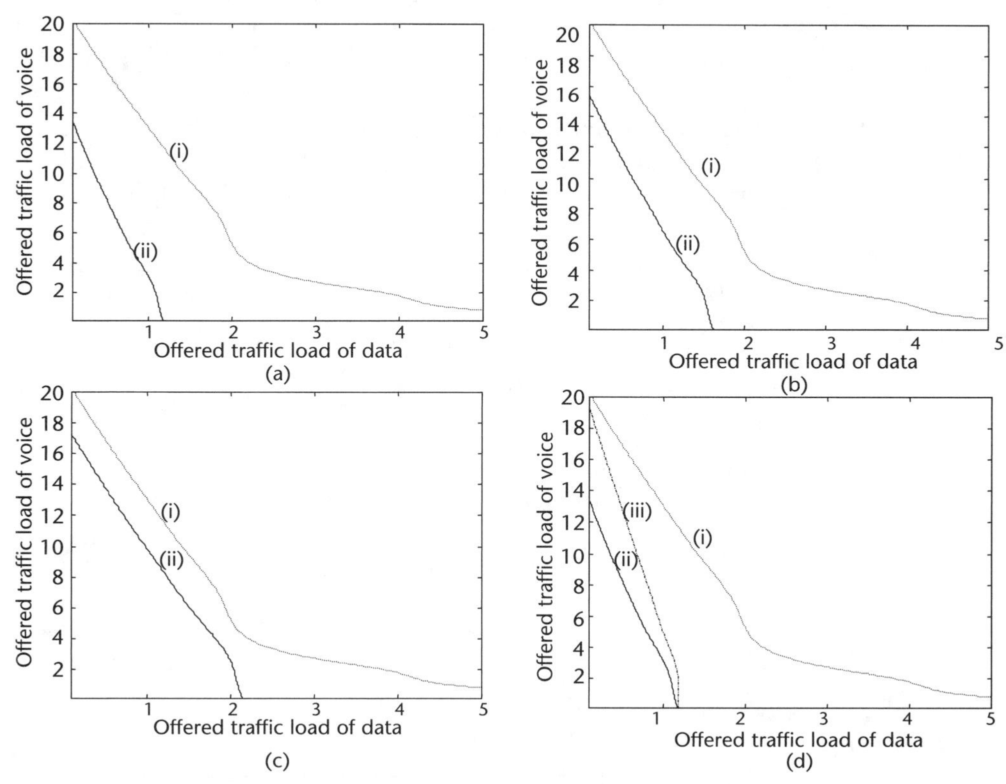

Figure 9.2 Erlang capacity for the different GoS requirements with $N = 100$: (a) when the voice GoS is 2% and the data GoS is 2%, (b) when the voice GoS is 2% and the data GoS is 5%, (c) when the voice GoS is 2% and the data GoS is 10%, and (d) when the voice GoS is 2%, the data GoS is 2%, and the average GoS is 2%. For each case, the curve represented by (i) is the Erlang capacity with respect to voice traffic, the curve represented by (ii) is the Erlang capacity with respect to data traffic, and the curve represented by (iii) is the Erlang capacity with respect to average GoS.

traffic. That is, the call blocking probabilities experienced by voice and data traffic are always less than 2% within the Erlang capacity region.

However, low Erlang capacity is archived, and the call blocking experienced by the voice call is relatively good, as compared with the required GoS—see Figure 9.2(a). Second, we consider the case where the different GoS requirements are given; a 2% GoS requirement is given for voice traffic, and 5% and 10% GoSs are given for data traffic, respectively. Figure 9.2(b, c) shows that the higher the GoS requirement of data is, the larger the Erlang capacity will be. However, the call blocking probabilities experienced by data traffic will be increased up to 5% and 10%, respectively, with heavy traffic load. That is, the Erlang capacity can be expanded at the price of the deteriorated GoS of data. Finally, as an alternative way, let's consider the Erlang capacity with respect to the average call blocking probability to combat the unbalanced call blocking probabilities between voice and data. We define the average call blocking probability as $P_{(blocking,\ ave)} = (\rho_v \cdot P_{(blocking,\ voice)} + \chi \cdot \rho_d \cdot P_{(blocking,\ data)})/(\rho_v + \chi \cdot \rho_d)$ where $1 \leq \chi \leq \Lambda$.

Figure 9.2(d) shows that this approach allows the Erlang capacity to be enhanced when the average call blocking probability is within about 2%. Additionally, the parameter χ can be used as a weighting factor (i.e., as χ is closer to Λ, more weight is given to data call blocking). Figure 9.3 shows the effect of on the Erlang capacity. These cases, so far mentioned, may be more suitable to the initial stages of the data service offering, where a service provider allows data call blocking to be higher than voice call blocking so that data traffic does not have a significant impact on voice traffic.

It is intuitive that the more CEs there are, the larger the Erlang capacity will be. The Erlang capacity, however, will be saturated after a certain value of CEs due to the insufficient channels per sector. For deeper consideration of the effect of CEs on

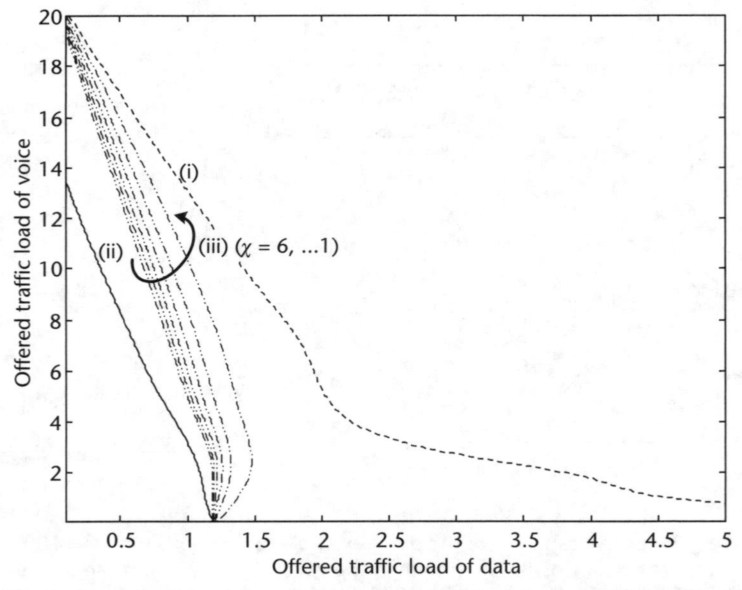

Figure 9.3 Effect of χ on the Erlang capacity with $N = 100$ when the voice GoS is 2%, the data GoS is 2%, and the average GoS is 2%. The curve represented by (i) is the Erlang capacity with respect to voice traffic, the curve represented by (ii) is the Erlang capacity with respect to data traffic, and the curve represented by (iii) is the Erlang capacity with respect to average GoS.

Erlang capacity, we assume that the offered load of data is proportional to that of voice, whereby the dimension of Erlang capacity can be reduced into to 1. Let $p(\equiv \rho_d/\rho_v)$ be the traffic ratio of data to voice. Figure 9.4 shows Erlang capacity as a function of the number of CEs with the different values of p (p = 1%, 2%, and 5%). From Figure 9.4, we observe that the more p there is, the less Erlang capacity there is (i.e., the introduction of more data traffic causes the Erlang capacity to be reduced).

In addition, we observe that the Erlang capacity region can be divided into three regions. In the first region, up to around 60 CEs, Erlang capacity increases linearly with the increase of the CEs. This means that call blocking, in this region, occurs mainly due to the limitation of CEs in the BS. In the second region, between about 60 CEs and 70CEs, Erlang capacity is determined by the interplay between the limitation of CEs in the BS and the insufficient channels per sector. Finally, in the third region, with more than 70 CEs, Erlang capacity is saturated, and call blocking is mainly caused by insufficient channels per sector. In particular, Figure 9.4 can be utilized to select the proper number of CEs in the BS that are required to accommodate the given traffic loads of voice and data. For example, if there is a voice traffic load of 8 Erlang and data traffic load of 0.08 Erlang per sector, respectively, which corresponds to p = 1%, there might be a question of how many CEs are needed to support these traffic loads. To answer this question, we recommend using more than 46 CEs in the BS, based on Figure 9.4.

9.4.2 Case of Multiple FAs and Graphic Interpretation Method

Until now, we have only considered one CDMA carrier. In order to meet a higher capacity requirement, multiple CDMA carriers are utilized, which are called multi-FA systems. In multi-FA systems, when a CDMA system carrier is licensed a

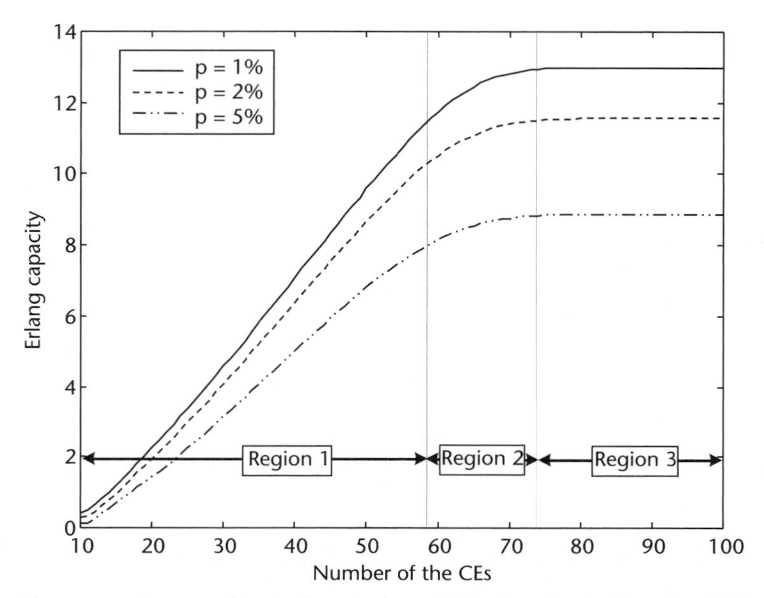

Figure 9.4 Erlang capacity according to the number of CEs when the traffic ratio of data to voice p is 1%, 2%, and 5%, respectively.

dedicated spectrum bandwidth, the total bandwidth is separated into a certain number of contiguous frequency allocations, and each subband facilitates a separate *narrowband* CDMA system. In this chapter, it is assumed that each FA subband has the 1.25-MHz frequency bandwidth. Additionally, the system performance of the multi-FA systems could be varied with the channel assignment methods among the multiple CDMA carriers. Here, we consider the CCCA method as a channel assignment method. The CCCA scheme combines all traffic channels in a system. When a BS receives a new call request, it searches the least occupied CDMA carrier and allocates a traffic channel in that carrier (i.e., arrivals of call attempts in a CDMA carrier are dependent upon the status of other CDMA carriers' occupation).

Conceptually, the multi-FA CDMA systems with P CDMA carriers under the CCCA scheme support $\hat{C}_{ETC} \cdot P$ basic channels per sector if each CDMA carrier provides $\hat{C}_{ETC}$ basic channels, where P is used for representing the number of the used CDMA carriers [10]. Then, similarly to the case of one CDMA carrier, which corresponds to the one-FA system, the Erlang capacity of the multi-FA systems with P CDMA carriers under the CCCA scheme can be evaluated by replacing $\hat{C}_{ETC}$ with $\hat{C}_{ETC} \cdot P$ and using the analytical procedures presented in Section 9.3.

Figure 9.5 shows Erlang capacity as a function of CEs according to the number of CDMA carriers when the traffic ratio of data to voice, p, is 1%. From Figure 9.5, the following observations can be made:

- The saturation values of the Erlang capacity according to the number of used CDMA carriers, which are denoted by ⇑ in Figure 9.5, have a linear property.
- For each number of the used CDMA carriers, the Erlang capacity according to the number of CEs also has a linear property at the first region.

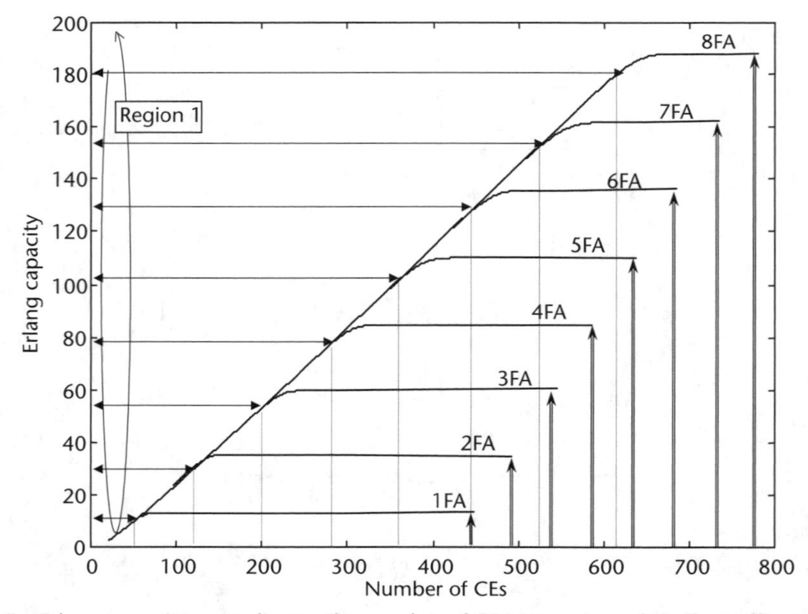

Figure 9.5 Erlang capacity according to the number of CDMA carriers when the traffic ratio of data to voice, p, is 1%.

The main advantage of these properties is that we can estimate the Erlang capacities for high FAs (e.g., five, six, seven, and eight) by using the linear regression with the Erlang capacity results of low FAs (e.g., one, two, three, and four). This linear regression approach for evaluating Erlang capacity is very attractive to traffic engineers, especially when we calculate the Erlang capacity for CDMA systems with a high FA. The reason is as follows: For the evaluation of Erlang capacity, the call blocking probability experienced by each call should be calculated. In the case of the analytical method proposed in this chapter, the following calculation amount is required for calculating the call blocking probability of each call:

$$\left\lfloor \frac{\hat{C}_{ETC}}{\Lambda} \right\rfloor \cdot \hat{C}_{ETC} \cdot P^2 + \left(\left\lfloor \frac{\hat{C}_{ETC}}{\Lambda} \right\rfloor \cdot \hat{C}_{ETC} \right)^3 \cdot P^6 + \left(\left\lfloor \frac{\hat{C}_{ETC}}{\Lambda} \right\rfloor \cdot \hat{C}_{ETC} \right)^3 \cdot P^6 \qquad (9.17)$$

The first term of (9.17) is for the calculation of sector state probability [see (9.3)], the second term is for the calculation of C [see (9.11)], and the last term is required for finding the call blocking states [see (9.12) and (9.13)].

Here, note that the complexity degree of the proposed method is increased proportionally to the sixth power of the number of used CDMA carriers. Subsequently, it is impractical to calculate Erlang capacity according to the numerical procedures presented in Section 9.3, especially when the number of the used CDMA carriers, P, is larger than four. For these cases, we suggest using the estimation method rather than direct numerical analysis. In addition, in Chapter 10, we will suggest an approximate analysis method for calculating Erlang capacity for CDMA systems with multiple sectors and multiple FA bands.

In order to estimate the Erlang capacity for a given FA (especially high FA) with the Erlang capacity results of low FAs, first we estimate the saturation value of Erlang capacity for a given FA. In the second phase, we estimate the slope of the Erlang capacity for a given FA. Finally, in the last phase, we estimate the Erlang capacity as a function of CEs for a given FA by combining the slope estimation and the saturation value estimation of Erlang capacity. Here, we consider the 7-FA case to illustrate the estimation procedures.

9.4.2.1 Saturation Value Estimation of Erlang Capacity for a High FA

As a result of Erlang capacity analysis, it is observed that the saturation value of Erlang capacity is determined by the number of used CDMA carriers where the call blocking is mainly caused by the limit of traffic channels per sector.

The vertical arrows in Figure 9.5, $\Uparrow$, represent the heights of the saturation values of Erlang capacity according to the number of used CDMA carriers when p is 1%. Figure 9.5 shows that the saturation values of Erlang capacity according to the number of used CDMA carriers have a linear property.

With this observation, let's estimate the saturation values of Erlang capacity for a high FA through the linear regression of those for low FAs. From the saturation values of Erlang capacity for low FAs ($P = 1, \ldots, 4$), where there are n points, $\mathbf{x}_i$, $i = 1, 2, \ldots, n$ (e.g., $n = 4$) with each $\mathbf{x}_i = [x_i, y_i]^T$ in which x_i is the number of used FAs and y_i is the saturation value of Erlang capacity corresponding to x_i, it would appear that we can approximately fit a line of the form

$$y_i \approx ax_i + b \tag{9.18}$$

for suitably chosen slope a and intercept b.

According to [11], the best estimation of a and b in the weighted least-squares sense is given as

$$\begin{bmatrix} a \\ b \end{bmatrix} = \left(A^H W A\right)^{-1} A^H W y \tag{9.19}$$

where $A = \begin{bmatrix} x_1 & x_2 & \dots & x_n \\ 1 & 1 & \dots & 1 \end{bmatrix}^T$, W is a weighting matrix reflecting the confidence in the data, and $\mathbf{y} = [y_1, y_2, \dots, y_n]^T$. In this chapter, we select a practical vector of confidence, W, as diag$\{10^1, 10^2, \dots, 10^n\}$ to incorporate the degree of data confidence increasing with the increment of the index of x_i, while $W = I$ corresponds to the estimation in the sense of regular least squares, where I is a unit matrix. Here, it is noteworthy that even though we select weighting matrix W somewhat intuitively, the other forms of weighting matrix W may be adopted for the better estimation of a and b.

Finally, we can estimate the saturation value of Erlang capacity for a high FA by using a linear equation, (9.18). Figure 9.6 illustrates the saturation values of Erlang capacity for high FAs (5 FA–8 FA) that are estimated from those of low FAs (1 FA–4 FA), and the saturation values of Erlang capacity that are calculated from the analytical procedure. The calculated Erlang capacity is plotted with "O" the estimated Erlang capacity with respect to least squares with "□" and the estimated Erlang capacity with respect to weighted least squares with "*". The estimated Erlang capacity is quite close to the calculated Erlang capacity, though further improvement may be possible.

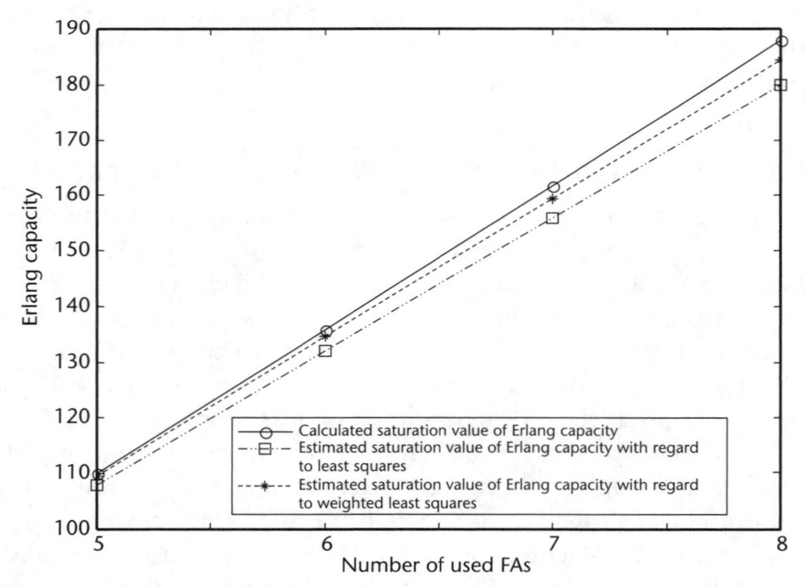

Figure 9.6 Estimated saturation values of Erlang capacity with saturation values of Erlang capacity for 1 FA–4 FA cases.

9.4.2.2 Slope Estimation of Erlang Capacity for a High FA

As with Figure 9.5, the Erlang capacity versus the number of CEs has a linear property at the first region for each number of CDMA carriers. Similarly to the case of the estimation of saturation values of Erlang capacity, the Erlang capacity according to the number of CEs can be estimated with the Erlang capacity results of low FAs. Figure 9.7 shows the slopes of the Erlang capacity for 7 FA that are estimated with the Erlang capacity results of 1 FA, 2 FA, 3 FA, and 4 FA, respectively.

From Figure 9.7, it is observed that the estimated slope of the Erlang capacity is closer to the calculated slope of Erlang capacity when the Erlang capacity results for the higher FA are utilized for the estimation process. Furthermore, it is observed that we have to analyze the Erlang capacity at least up to 3 FA, and then estimate the slope of Erlang capacity for 7 FA with those Erlang capacity results in order to properly estimate the slope of Erlang capacity for 7 FA.

9.4.2.3 Estimation of Erlang Capacity

Figure 9.8 shows the estimated Erlang capacity for 7 FA as a function of the number of CEs, which are obtained through the combination of the slope estimation and the saturation value estimation of Erlang capacity. Figure 9.9 shows the estimation errors for 7 FA between the calculated Erlang capacity and the estimated Erlang capacity. From Figure 9.9, it is observed that we can estimate Erlang capacity for 7 FA within the estimated error of 2% with only the Erlang capacity results of 4 FA.

9.5 Conclusion

In this chapter, we presented an analytical procedure for the evaluation of Erlang capacity in the reverse link of the multimedia CDMA systems, by considering soft

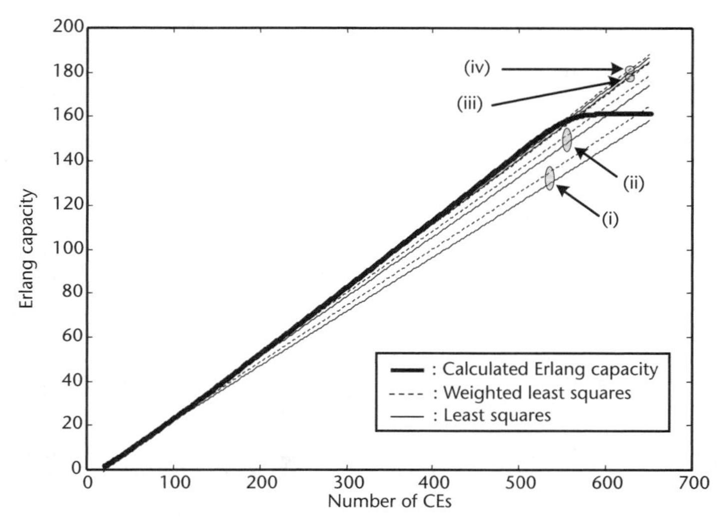

Figure 9.7 Estimated slopes of Erlang capacity for 7 FA; the curves represented by (i), (ii), (iii), and (iv) are estimated with the Erlang capacity results of 1 FA, 2 FA, 3 FA, and 4 FA, respectively. For each case, the solid line is estimated with regard to least squares, and the dotted line is estimated with regard to weighted least squares.

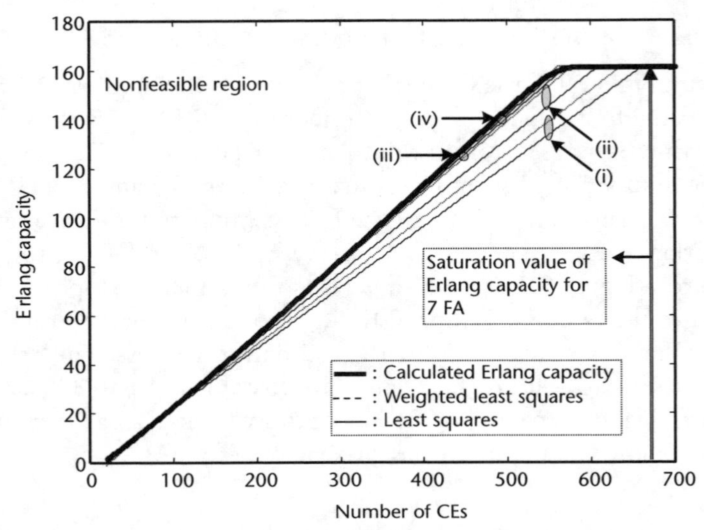

Figure 9.8 Estimated Erlang capacity through slope estimation and the estimation of saturation value of Erlang capacity; the curves represented by (i), (ii), (iii), and (iv) are estimated with the Erlang capacity results of 1 FA, 2 FA, 3 FA, and 4FA, respectively. For each case, the solid line is estimated with regard to least squares, and the dotted line is estimated with regard to weighted least squares.

blocking as well as hard blocking. For the performance analysis, a multidimensional Markov chain is developed. Through a numerical example of the voice/data CDMA system, we observe that data users have more impact on the Erlang capacity than voice users do. It is observed that the Erlang capacities with respect to all traffic should be balanced to enhance total system Erlang capacity.

To get this tradeoff, we allocate the different GoS requirements for voice and data traffic and observe the effect of the different GoS requirements on Erlang

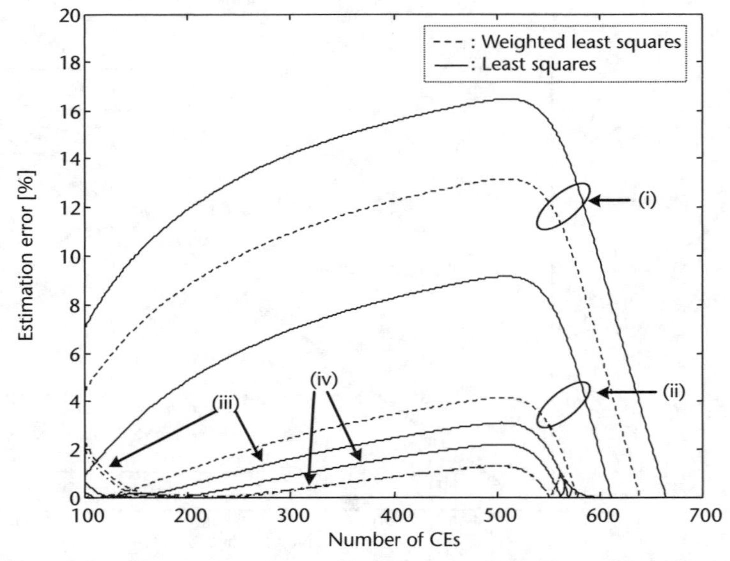

Figure 9.9 Estimation error of Erlang capacity for 7 FA; the curves represented by (i), (ii), (iii), and (iv) come from the estimation with the Erlang capacity results of 1 FA, 2 FA, 3 FA, and 4 FA, respectively. For each case, the solid line is estimation error with least squares and the dotted line is estimation error with weighted least squares.

capacity. In addition, the effect of the CEs on Erlang capacity is investigated, and it is found out that the more CEs there are, the larger Erlang capacity will be. However, the Erlang capacity is saturated after a certain value of CEs, where call blocking is mainly caused by insufficient channels per sector. Furthermore, we expand our approach to consider the multi-FA systems that support multiple CDMA carries, where Erlang capacity is almost impractical to be numerically analyzed. For high-FA cases, the graph interpretation method is suggested, and it is observed that Erlang capacity for a high FA can be well estimated through linear regression with the Erlang capacity results of low FAs. Finally, it is expected that the Erlang capacity analysis method can be utilized mainly in two ways. For given loads of voice and data traffic, it can be used for selecting the appropriate values of system operating parameters to support given traffic loads with QoS and GoS requirements, or it can be used for estimating the supportable size of the system for given system parameters.

References

[1] Kim, K. I., *Handbook of CDMA System Design, Engineering and Optimization,* Englewood Cliffs, NJ: Prentice Hall, 2000.

[2] Sampath, A., P. S. Kumar, and J. M. Holtzman, "Power Control and Resource Management for a Multimedia CDMA Wireless System," *IEEE Proc. of International Symposium on Personal, Indoor, and Mobile Radio Communications*, 1995, pp. 21–25.

[3] Yang, Y. R., et al., "Capacity Plane of CDMA System for Multimedia Traffic," *IEEE Electronics Letters*, 1997, pp. 1432–1433.

[4] Koo, I., et al., "A Generalized Capacity Formula for the Multimedia DS-CDMA System," *IEEE Proc. of Asia-Pacific Conference on Communications*, 1997, pp. 46–50.

[5] Viterbi, A. M., and A. J. Viterbi, "Erlang Capacity of a Power-Controlled CDMA System," *IEEE Journal on Selected Areas in Communications,* 1993, pp. 892–900.

[6] Sampath, A., N. B. Mandayam, and J. M. Holtzman, "Erlang Capacity of a Power Controlled Integrated Voice and Data CDMA System," *IEEE Proc. of Vehicular Technology Conference*, 1997, pp. 1557–1561.

[7] Matragi, W., and S. Nanda, "Capacity Analysis of an Integrated Voice and Data CDMA System," *IEEE Proc. of Vehicular Technology Conference*, 1999, pp. 1658–1663.

[8] Koo, I., et al., "Analysis of Erlang Capacity for the Multimedia DS-CDMA Systems," *IEICE Trans. Fundamentals*, 1999, pp. 849–855.

[9] Kelly, F., "Loss Networks," *The Annals of Applied Probability*, 1991, pp. 319–378.

[10] Song, B., J. Kim, and S. Oh, "Performance Analysis of Channel Assignment Methods for Multiple Carrier CDMA Cellular Systems," *IEEE Proc. of VTC (Spring),* 1999, pp. 10–14.

[11] Moon, T. K., and W. C. Stirling, *Mathematical Methods and Algorithms*, Englewood Cliffs, NJ: Prentice Hall, 2000.

CHAPTER 10

Approximate Analysis Method for CDMA Systems with Multiple Sectors and Multiple FAs

The analytic methods shown in the previous chapters and in [1] require tedious calculations for call blocking and Erlang capacity, especially when the system supports multiple sectors and multiple FA bands. In this chapter, we propose an approximate analysis method, reducing the exponential complexity of the old method [1] down to the linear complexity. The approximated results also provide a difference of only a few percent from the exact value [1], which makes the proposed method practically useful.

10.1 Introduction

In a CDMA system, call attempts may be blocked due not only to the scarcity of CEs in the BS but also to an excess of the maximum number of concurrent users. The CE is a hardware element that performs the baseband signal processing of the received DS signal for a given channel in the BS and is practically an important system resource. Call blocking, which is caused by insufficient CEs in the BS, is called hard blocking. In addition, the excessive interference due to concurrent users causes call blocking, which is called soft blocking and occurs typically when the number of active users exceeds the maximum number of concurrent users.

At a sectorized CDMA cell, CEs in the BS are pooled and can be assigned to any user regardless of the sector. It would be wasteful to provide CEs per sector based on the per-sector traffic loads, because the trunking efficiency gained by pooling the CEs would be lost. In such case, it is very important for traffic engineers to determine the proper number of CEs in the BS with which the call blocking probability meets the required call blocking requirement. As a relevant research work, Kim in [1] and Chapter 9 presented a method that computers the call blocking probability. It determines the required number of CEs in the BS, based on the individual traffic loads of the sectors. The methods proposed in [1] and Chapter 9, however, require a great deal of calculation for computing the call blocking probability, especially when CEs are shared across more than three sectors (i.e., a multisector system). In addition, practical CDMA systems utilize multiple CDMA carriers to accommodate continuously increasing CDMA subscribers. In such multiple CDMA carriers case, the calculation complexity of the method proposed in [1] for computing the call blocking probability is increased proportionally to the Kth power of the number of

CDMA carriers when CDMA systems support single service traffic with K sector cells.

In this chapter, we propose an approximate method to efficiently compute call blocking probability for CDMA systems with the multiple sectors and multiple frequency allocation bands. The proposed approximate method shows similar results to those of [1] while reducing the exponential complexity of the old method [2] down to the linear complexity.

10.2 System Model

To compute the call blocking probability, let's consider a CDMA system modeled as follows:

- We consider a multicell CDMA system with K sectors supporting P multiple CDMA carriers, where K and P denote the number of sectors and the number of CDMA carriers, respectively.
- At each sector, each CDMA carrier facilitates a narrowband CDMA system whose signals employ DS spreading and are transmitted in one CDMA carrier. For each CDMA carrier, although there is no hard limit on the number of mobile users served, there is a practical limit on the number of concurrent users to control the interference between users that have the same pilot signal. The maximum number of concurrent users that a CDMA carrier can support with QoS requirements, such as data transmission rate and the required E_b/N_0, was found, based on the maximum tolerable interference [2, 3].
- We assume that each CDMA carrier has an M user limit per sector. In addition, it is assumed that the CCCA scheme is used as a channel assignment method among the multiple CDMA carriers. Under the CCCA scheme, a CDMA system with P multiple CDMA carriers has $M \cdot P$ user limits per sector [4].
- There are N CEs at each BS, where all CEs are pooled for efficient usage such that any CE can be assigned to any user in the cell, regardless of sector.
- The traffic impinging on a cell is assumed to be characterized by Poisson arrivals and exponentially distributed holding times. If λ denotes the arrival rate of calls in a region and $1/\mu$ denotes the average holding time, then the traffic load is given as $\rho = \lambda/\mu$. The traffic load for the K sectors will be denoted $(\rho_1, \rho_2, \ldots, \rho_K)$, where $\rho_i = \lambda_i / \mu_i$ $(i = 1, 2, \ldots, K)$.

10.3 Approximate Analysis Method

Each user shares the system resources with other users and competes with other users for the use of the system resources. In this situation, a call attempt may be soft blocked at each sector or be hard blocked in the BS. That is, in order for a call attempt to get service in a cell, the soft blocking of the call should not occur in a sector, and the hard blocking of the call should not occur in the BS. In this section, we define the soft blocking probability in sector i as $b_{(soft,i)}$ and the hard blocking probability in the BS as $b_{(hard)}$. We then present an approximate method for efficiently computing the call blocking probability.

As a result of Chapter 9, it is observed that the Erlang capacity region can be divided into three regions according to the number of CEs. In the first region, Erlang capacity increases linearly with the increase of the CEs, which means that call blocking, in this linear region, occurs mainly due to the limitation of CEs in the BS. In the second region, Erlang capacity is determined by the interplay between the limitation of CEs in the BS and insufficient traffic channels per sector. Finally, Erlang capacity is saturated in the third region, where call blocking is mainly due to user limit per sector. Because the first and third regions are dominant among the three regions, in the proposed analysis method, we intuitively decouple the calculation stages of soft blocking and hard blocking for the simplicity of computation, by which the soft-blocking and hard-blocking probabilities can be separable as a closed-form equation, respectively. However, these closed-form equations may not provide universal values of soft-blocking and hard-blocking probabilities because practically the soft-blocking and hard-blocking probabilities affect each other in the blocking model being considered.

In order to consider mutual effects between the hard blocking in the BS and the soft blocking in each sector, in our manuscript we introduce *coupling* parameters, $\bar{\rho}_i$ and α.

First, let's consider the closed-form equation for the soft blocking probability. Because CDMA systems with P multiple CDMA carriers support $M \cdot P$ users per sector without any QoS degradation, we assume that the blocked calls are cleared and that the maximum number of supportable users in a sector is $M \cdot P$, respective of loading. Then, given the sector traffic load, the probability of having exactly n users in sector i, $\pi_{(i,n)}$, becomes [5]:

$$\pi_{(i,n)} = \frac{\frac{\bar{\rho}_i^n}{n!}}{\sum_{k=0}^{M \cdot P} \frac{\bar{\rho}_i^k}{k!}} \quad n = 0, \ldots, M \cdot P \text{ and } i = 1, 2, \ldots, K \tag{10.1}$$

where $\bar{\rho}_i$ is defined to consider the traffic load of the ith sector that is somewhat reduced due to the limitation of CEs in the BS. Note that when n becomes $M \cdot P$, $\pi_{(i, M \cdot P)}$ is equivalent to the blocking probability according to Erlang B.

Then, the closed-form equation for the soft-blocking probability in sector i is given by

$$b_{(soft,i)} = \frac{\frac{\bar{\rho}_i^{M \cdot P}}{M \cdot P!}}{\sum_{k=0}^{M \cdot P} \frac{\bar{\rho}_i^k}{k!}} \quad i = 1, 2, \ldots, K \tag{10.2}$$

In order for the calls, which are not soft blocked in each sector, to get the services, there should be sufficient CEs in the BS to support those calls. If there are not sufficient CEs in the BS, those calls will be hard blocked. Because all CEs available in the BS are pooled and assigned to any call regardless of sectors, α is introduced to consider the traffic load that is offered to the BS from each sector and is defined as $\alpha = \sum_{i=1}^{K} \rho_i \cdot \left(1 - b_{(soft,i)}\right)$.

Then, when there are N CEs in the BS, similarly to the soft-blocking case, the closed-form equation for the hard-blocking probability is given as

$$b_{(hard)} = \frac{\dfrac{\alpha^N}{N!}}{\sum_{k=0}^{N} \dfrac{\alpha^k}{k!}} \tag{10.3}$$

Subsequently, the problem to evaluate the soft-blocking and hard-blocking probabilities is to solve (10.2) and (10.3), which are mutually linked by two coupling parameters $\overline{\rho}_i$ and α. For the calculation of these blocking probabilities, in this chapter, we propose an iteration method, which is described in Figure 10.1. Let's let $b_{(soft,\,i)}(m)$ and $b_{(hard)}(m)$ be the value of $b_{(soft,\,i)}$ and $b_{(hard)}$ at the mth iteration for m = 1, 2, 3, ..., respectively, and let $b_{(soft,\,i)}(0)$ and $b_{(hard)}(0)$ be the initial value for the recursion. At the mth iteration, $b_{(soft,\,i)}(m)$ is computed using (10.2) with $\overline{\rho}_i = \rho_i \cdot (1 - b_{(hard)}(m))$, where we intuitively let $\overline{\rho}_i$ be $\rho_i\ (1 - b_{(hard)}(m))$ to reflect on the effect of the limited number of CEs (N) in the BS on the soft-blocking probability in the ith sector through the feedback quantity of $b_{(hard)}(m)$. Also, at the mth iteration, $b_{(hard)}(m)$ is computed using (10.3) with $\alpha = \sum_{i=1}^{K} \rho_i \cdot \left(1 - b_{(soft,i)}(m-1)\right)$, where we also intuitively let α be $\sum_{i=1}^{K} \rho_i \cdot \left(1 - b_{(soft,i)}(m-1)\right)$ to consider the effect of the user limit ($M \cdot P$)

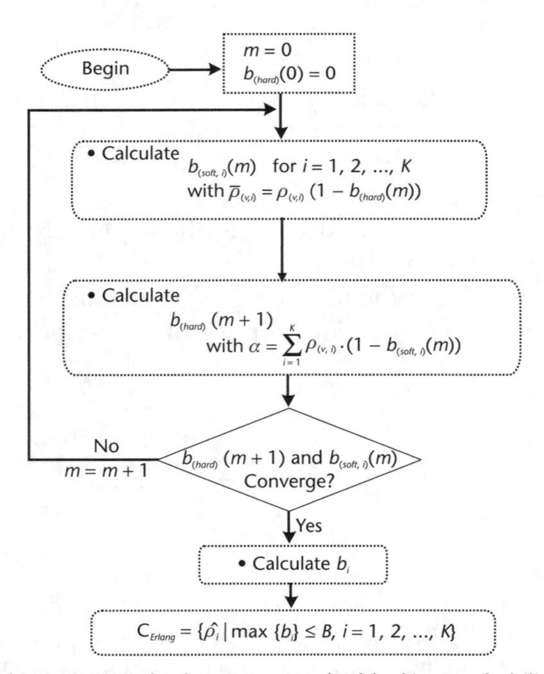

Figure 10.1 Proposed iteration method to compute the blocking probability.

in each sector with the quantity of $b_{(soft,\, i)}(m-1)$ and the traffic loads of each sector $(\rho_1, \rho_2, \ldots, \rho_K)$ on the hard-blocking probability in the BS.

Here, it is noteworthy that even though we select the coupling parameters ρ_i and α somewhat intuitively, the other forms of the coupling parameters may be adopted for the better calculation of the soft-blocking and hard-blocking probabilities. Then, the iteration procedure takes the following steps:

1. Define $m = 0$, and set $b_{(hard)}(0)$.
2. Compute $b_{(soft,\, i)}(m)$ with $\overline{\rho}_i = \rho_i \cdot (1 - b_{(hard)}(m))$ using (10.2) for all i ($i = 1, 2, \ldots, K$).
3. Compute $b_{(hard)}(m+1)$ with $\alpha = \sum_{i=1}^{K} \rho_i \cdot (1 - b_{(soft,\, i)}(m))$ using (10.3).
4. If $(|b_{(hard)}(m+1) - b_{(hard)}(m)|/b_{(hard)}(m+1)) < \tau$ (tolerance parameter), then stop the recursion. Otherwise, set $m = m + 1$ and go back to step 2.

From our numerical experiences, it is observed that this recursion always converges within a few iterations (generally less than five). Finally, the call blocking probability of the ith sector, b_i, is given as (10.4) for the convergence values of the soft-blocking and hard-blocking probabilities.

$$
\begin{aligned}
b_i &= 1 - \left(1 - b_{(soft,i)}\right) \cdot \left(1 - b_{(hard)}\right) \\
&= \frac{\dfrac{\overline{\rho}_i^{\,M \cdot P}}{M \cdot P!}}{\sum_{k=0}^{M \cdot P} \dfrac{\overline{\rho}_i^{\,k}}{k!}} + \frac{\dfrac{\alpha^N}{N!}}{\sum_{k=0}^{N} \dfrac{\alpha^k}{k!}} - \frac{\dfrac{\overline{\rho}_i^{\,M \cdot P}}{M \cdot P!}}{\sum_{k=0}^{M \cdot P} \dfrac{\overline{\rho}_i^{\,k}}{k!}} \cdot \frac{\dfrac{\alpha^N}{N!}}{\sum_{k=0}^{N} \dfrac{\alpha^k}{k!}}
\end{aligned} \tag{10.4}
$$

The problem of providing CEs for a CDMA BS having traffic loads $(\rho_1, \rho_2, \ldots, \rho_K)$ in the sectors reduces to getting the smallest number for which the blocking probability $(b_1, b_2, \ldots, b_K)$ meets the blocking requirement. Generally, the objective might involve the most heavily loaded sector, as in [1]:

$$
\max\{b_i\} < B \tag{10.5}
$$

where B is the required call blocking probability. To consider such an objective, we introduce the Erlang capacity, defined as the maximum traffic load of the most heavily loaded sector in which the blocking probability $(b_1, b_2, \ldots, b_K)$ meets the blocking requirement with a proper number of CEs such that

$$
C_{Erlang} = \left\{\hat{\rho}_i \,|\, b_{\hat{i}} \le B\right\} \text{ where } \hat{i} = \arg\max_i \{b_i\}, i = 1, 2, \ldots, K \tag{10.6}
$$

10.4 Calculation Complexity of the Proposed Method

In this section, we evaluate the calculation complexity of the approximate method, and compare it with that in [1] to illustrate the calculation efficiency of the approximate method. To do this, first we define O_1 as the calculation amount required to compute the call blocking probability in the single-sector case with the user limit of $M \cdot P$, and we define O_2 as the calculation amount required for solving the $M \cdot P$ linear simultaneous equations involving $M \cdot P$ variables. Typically, the call blocking probability for the case of single sector cells can be computed from the Erlang B model, which requires approximately $4M \cdot P$ multiplication operations [i.e., $O_1 \approx O(4\, M \cdot P)$]. Likewise, to solve $M \cdot P$ simultaneous equations with $M \cdot P$ variables, we need multiplication operations approximately on the order of $(M \cdot P)^3$ [i.e., $O_2 \approx O((M \cdot P)^3)$].

In [1], a marginal probability is introduced that allows derivation of the state probabilities in order to compute the blocking probability for the case of K sector cells. Note that the problem of finding the marginal probability of a given sector in a K-sector BS can be described equivalently to that of finding the eigenvector of a product of two matrixes corresponding to an eigenvalue equal to 1 [1]. The first matrix of the product is the conditional probability matrix of the sector under consideration when the sum of the number of CEs used by the other $(K - 1)$ sectors is given. Noting that the first matrix can be derived from the traditional Erlang B model, it necessitates the calculation amount of $(K - 1) \cdot (M \cdot P) \cdot O_1$. The second one is the conditional probability matrix on the number of CEs used by $(K - 1)$ sectors when given the number occupied by the sector under consideration, which can be derived from the $(K - 1)$ sector case. Considering that the second conditional probability can be found recursively, and each recursion requires the finding of the corresponding two conditional probabilities along with the solving of $(M \cdot P)$ simultaneous equations, there exists the following calculation amount to find the second conditional probability matrix.

$$\left((M \cdot P)^{K-1} + \sum_{i=1}^{K-2} i(M \cdot P)^{K-i}\right) \cdot O_1 + \left(\sum_{i=1}^{K-2} (M \cdot P)^i\right) \cdot O_2 \tag{10.7}$$

Consequently, the method proposed in [1] requires the following calculation complexity to find the marginal probability and further to compute the call blocking probability.

$$\begin{aligned} &\left\{\left((M \cdot P)^{K-1} + \textstyle\sum_{i=1}^{K-1} i(M \cdot P)^{K-i}\right) \cdot O_1 + \left(\textstyle\sum_{i=0}^{K-2} (M \cdot P)^i\right) \cdot O_2\right\} \\ &\approx \left\{2(M \cdot P)^{K-1} \cdot O_1 + (M \cdot P)^{K-2} \cdot O_2\right\} \end{aligned} \tag{10.8}$$

Additionally, it is noteworthy that the calculation complexity of the method in [1] increases exponentially (i.e., proportional to the Kth power of the number of multiple frequency allocations). Consequently, it could be computationally infeasible to do the calculation of the call blocking probability for the large multiple sector case or multiple frequency allocation case.

The approximate method decouples the calculation stages of soft blocking and hard blocking, which requires some iterations until converging to the satisfactory values. Let the iteration number be defined as γ, which is typically less than five from our numerical experience. At each iteration, we need the calculation amount of $K \cdot O_1$ for the computation of soft blocking in each sector and O_1 for the computation of hard blocking in the BS. Consequently, the approximate method requires just the calculation complexity of $\{(\gamma \cdot (K + 1) \cdot O_1)\}$ for computing the call blocking probability such that it reduces the exponential complexity of the old method [1] down to the linear complexity while providing approximated values that have a few percent difference with the exact values.

10.5 Numerical Example

In this section, assuming that the sectors are equally loaded, we provide calculated Erlang capacity per sector for the following cases:

- P CDMA carriers per sector (P = 1, 2, 3).
- Users limit per sector per carrier is 15.
- K-sector CDMA cells (K = 2, 3, 4).

Figure 10.2 shows the call blocking probabilities for diverse values of K when $P = 1$, $M = 15$, and $N = 35$. The dotted line indicates the call blocking probability computed according to the method in [1], and the solid line indicates the call blocking probability computed according to the approximate method. Figure 10.2 indicates a good match between the approximate value (solid line) and the exact value (dotted line). On the other hand, Table 10.1 shows the viewpoint of calculation complexity. In this case, the method of [1] requires approximately 6, 70, and 830 times the calculation amount of the approximate method when K = 2, 3, and 4, respectively.

Figure 10.3 shows the Erlang capacity per sector according to the number of the CEs for a diverse number of K (K = 2, 3, and 4), when $P = 1$, $M = 15$, and the call blocking requirement is given as 2%. The dotted and solid lines indicate Erlang capacities per sector that are calculated according to the method suggested in [1]

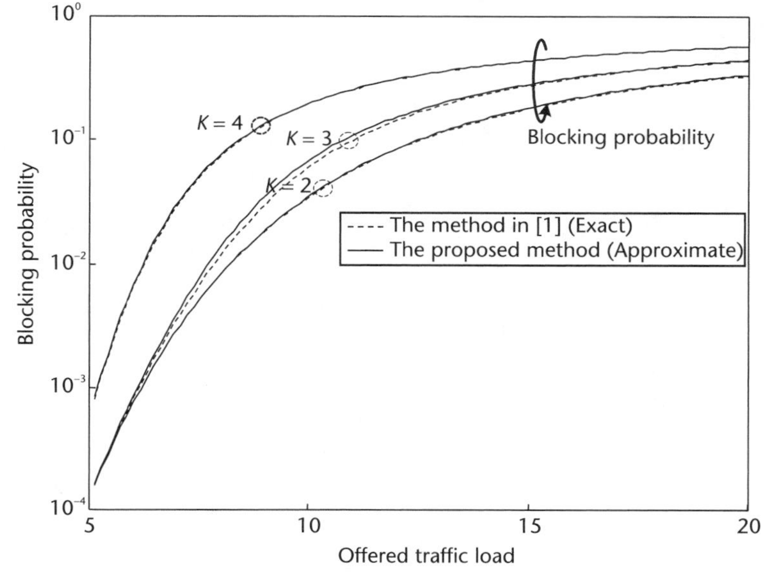

Figure 10.2 The call blocking probability for the diverse number of sectors (K), K = 2, 3, and 4, when $P = 1$, $M = 15$, and $N = 35$.

Table 10.1 Comparison of the Calculation Complexity of the Method in [1] and the Proposed Method

	Calculation Complexity		
K	*The Method in [1]*	*The Proposed Method*	*Complexity Ratio*
2	5.2×10^3	9×10^2	5.8
3	8.28×10^4	1.2×10^3	69
4	1.25×10^6	1.5×10^3	832

and the approximate method, respectively. From Figure 10.3, we observe that the approximate method provides similar results to those of [1]. Also, the dashed lines in Figure 10.3 indicate the Erlang capacity differences between the method suggested in [1] and the approximate method when $K = 2$, 3, and 4, respectively, which are smaller than 3% and decreases as the number of sectors increases.

Finally, Figure 10.4 shows the calculated Erlang capacity according to the number of CEs for different numbers of P ($P = 1, 2, 3$) when $K = 3$ and 2% call blocking objective is given. The dotted and solid lines indicate the Erlang capacities per sector that are calculated according to the method in [1] and the proposed approximate method, respectively. Figure 10.4 also indicates that the approximate method provides the similar results to those of [1], and the Erlang capacity differences for $P = 1, 2, 3$ are always smaller than 3% and decrease as the number of CDMA carriers increases.

10.5.1 An Interesting Observation: Two Traffic Parameters to Efficiently Approximate the Call Blocking Probability in CDMA Systems with Three Sectors

For CDMA systems with three sectors, in this section, we show that the call blocking probability and Erlang capacity can be characterized just with two traffic parameters (the traffic load of the most loaded sector and the sum of traffic loads of the

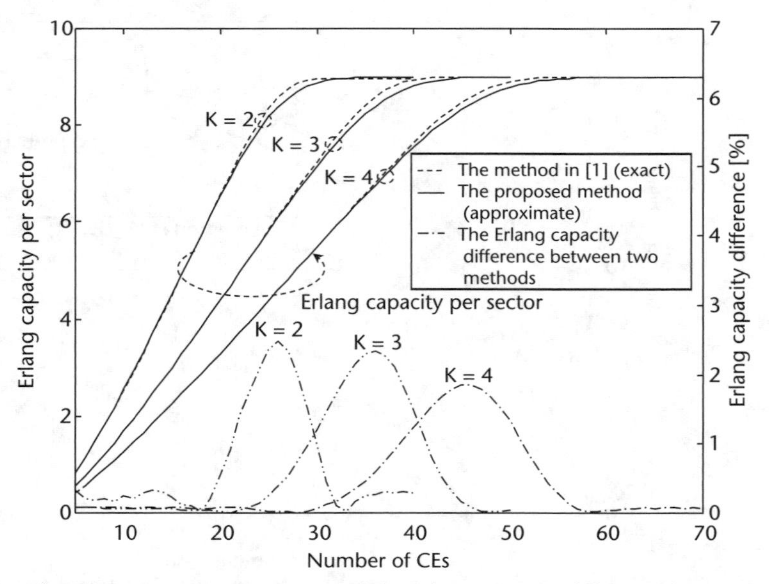

Figure 10.3 The Erlang capacity per sector and the Erlang capacity difference between the method suggested in [1] and the proposed approximate method for a diverse number of *K*.

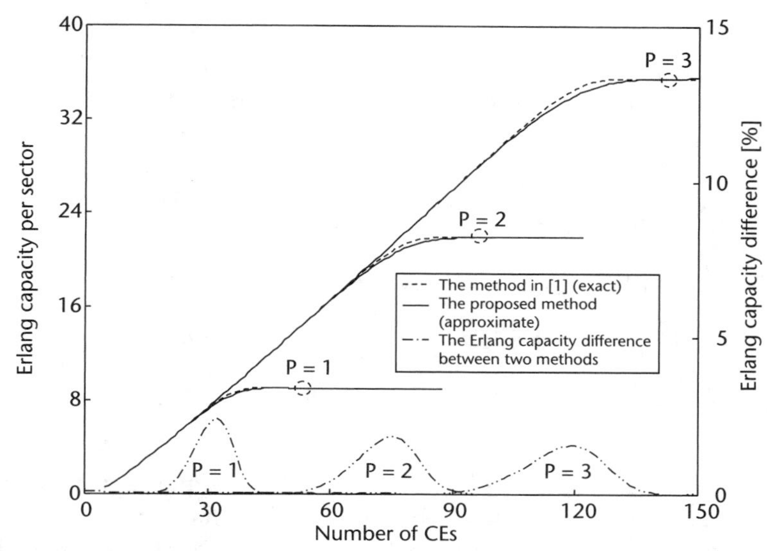

Figure 10.4 The Erlang capacity per sector and the Erlang capacity difference between the method suggested in [1] and the proposed approximate method for diverse number of *P*.

other remaining sectors) instead of three sector traffic loads, especially when the required call blocking probability is given less than $2e^{-2}$, which makes the traffic engineers manage the system more easily.

To explain this interesting observation, let's consider CDMA systems with three sectors employing the perfect directional antennas and assume that the traffic impinging on a cell is characterized by Poisson arrivals and exponentially distributed holding times. If λ denotes the arrival rate of calls in a region, and $1/\mu$ denotes the average holding time, then the traffic load is given as $\rho = \lambda/\mu$. The traffic load for three sectors will be denoted (ρ_1, ρ_2, ρ_3) where $\rho_i = \lambda_i/\mu_i$ $(i = 1, 2, 3)$. In order to consider unequal traffic load among three sectors, we introduce the sector traffic ratio, *JJ*, which is defined as following:

$$JJ = \frac{\text{mid}(\rho_1, \rho_2, \rho_3) + \min(\rho_1, \rho_2, \rho_3)}{\max(\rho_1, \rho_2, \rho_3)} \tag{10.9}$$

where "mid" function takes the middle one among three elements, "min" function takes the minimum one, and "max" function takes the maximum one. For the convenience of analysis, we assume that the first sector is the most loaded sector [i.e., $\rho_1 = \max(\rho_1, \rho_2, \rho_3)$]. Then, the traffic loads of the other remaining sectors, ρ_2 and ρ_3, can be reexpressed as follows:

$$\rho_2 = JJ \cdot \rho_1 \cdot (1 - p) \tag{10.10}$$

$$\rho_3 = JJ \cdot \rho_1 \cdot p \tag{10.11}$$

where p is a parameter that takes a typical value between $\max(0, 1 - 1/JJ)$ and $\min(1, 1/JJ)$. In the case of $p = 1/2$, ρ_2 and ρ_3 are identical. In addition, three sectors are equally loaded when $JJ = 2$ and $p = 1/2$. Subsequently, we can reexpress (ρ_1, ρ_2, ρ_3) into $(\rho_1, JJ \cdot \rho_1 \cdot (1 - p), JJ \cdot \rho_1 \cdot p)$. For the three-sector case, the coupling parameter α is given as $\alpha = \sum_{i=1}^{3} \rho_i \cdot \left(1 - b_{(soft,i)}\right)$. Further, it is noteworthy that (10.12) can be held in the practical range of the call blocking probability less than $2e^{-2}$.

$$
\begin{aligned}
\alpha &= \sum_{i=1}^{3} \rho_i \cdot \left(1 - b_{(soft,i)}\right) \\
&= \rho_1 - \rho_1 \cdot b_{(soft,1)} + JJ \cdot \rho_1 - JJ \cdot \rho_1 \cdot b_{(soft,2)} \cdot (1-p) - JJ \cdot \rho_1 \cdot b_{(soft,3)} \cdot p \\
&= \rho_1 \cdot \left(1 - b_{(soft,1)}\right) + JJ \cdot \rho_1 \cdot \left[1 - \left(b_{(soft,2)} - p \cdot \left(b_{(soft,2)} - b_{(soft,3)}\right)\right)\right] \\
&\approx \rho_1 \cdot \left(1 - b_{(soft,1)}\right) + JJ \cdot \rho_1
\end{aligned}
\tag{10.12}
$$

where $(b_{(soft,2)} - p \cdot (b_{(soft,2)} - b_{(soft,3)}))$ is negligible compared to 1 as long as the interesting range of the call blocking probability is less than $2e^{-2}$. From this observation, we know that the call blocking probability is nearly not affected by the traffic parameter p, and it is mainly dependent on two traffic parameters: the traffic load of the most loaded sector, ρ_i, and the sector traffic ratio, JJ, for the practical range of call blocking probabilities. From this observation, we conclude that it is sufficient for us to consider just two traffic parameters (ρ_1, JJ) instead of all traffic loads of three sectors (ρ_1, ρ_2, ρ_3) when calculating the call blocking probability, as long as the required call blocking probability is less than $2e^{-2}$.

Figure 10.5 shows the effect of traffic parameter p on the call blocking probability for diverse user limits (M=10, 12, or 15) when $JJ = 1.5, \rho_1 = 10$, and $N = 45$. These results are obtained by both the method shown in [1] and the proposed method, and they are same in that case. As seen in Figure 10.5, the call blocking probability is independent of the traffic parameter of p where $p \in (0.34, 0.667)$, which verifies

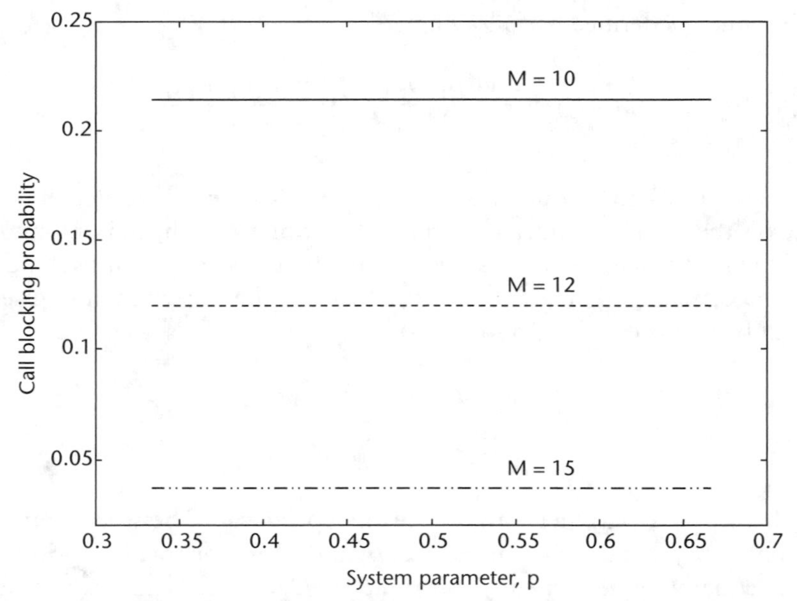

Figure 10.5 Effect of the traffic parameter p on the call blocking probability when $JJ = 1.5, \rho_1 = 10$, and $N = 45$.

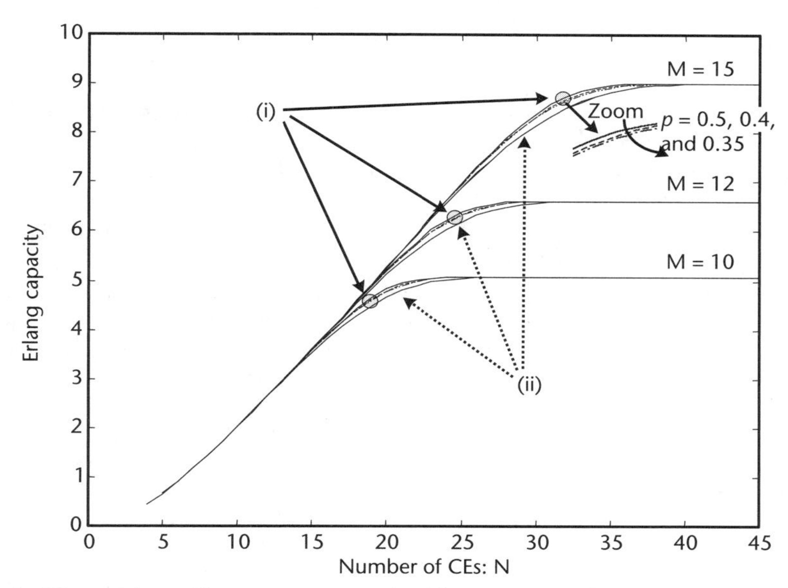

Figure 10.6 Effect of the traffic parameter p on the Erlang capacity when $JJ = 1.5$, and the call blocking objective is given as 2%.

that the call blocking probability can be characterized with just the two traffic parameters (sector traffic load, $JJ = 1.5$, and the traffic load of the most heavily loaded sector, $\rho_1 = 10$) instead of three sector traffic loads ((10, 15(1 − p), 15p)).

Figure 10.6 shows the Erlang capacity per sector as a function of CEs for diverse values of p ($p = 0.35$, 0.4, and 0.5). The plots denoted by (i) are the Erlang capacities that are calculated according to the method shown in [1], while the plots denoted by (ii) are those according to the proposed method. Figure 10.6 shows that the traffic parameter p has no effect on the Erlang capacity for both cases. Similarly to the case of the call blocking probability, it means that Erlang capacity can be characterized just with two traffic parameters, JJ and ρ_1 for the given conditions.

10.6 Conclusion

For CDMA systems with multiple sectors, we propose an approximate analysis method for efficiently computing of the call blocking probability and the Erlang capacity. The approximate method shows similar results to those of [1] in the practical call blocking probability range of 0.1% to 5% in which traffic engineers are mainly interested, while it reduces the calculation complexity. It is noteworthy that even though only the single-service case is considered here, the proposed approximate approach can be expanded to the multiclass services case.

For CDMA systems with three sectors, we also show that the call blocking probability and Erlang capacity can be characterized by two traffic parameters (the traffic load of the most loaded sector and the sum of traffic loads of the other remaining sectors) instead of three sector traffic loads, especially when the required call blocking probability given is less than $2e^{-2}$, which makes the traffic engineers manage the system more easily.

References

[1] Kim, K. I., *Handbook of CDMA System Design, Engineering and Optimization,* Englewood Cliffs, NJ: Prentice Hall, 2000.

[2] Gilhousen, K. S., et al., "On the Capacity of a Cellular CDMA System," *IEEE Trans. on Vehicular Technology*, 1991, pp. 303–312.

[3] Yang, J. R., et al., "Capacity Plane of CDMA System for Multimedia Traffic," *IEEE Electronics Letters*, 1997, pp. 1432–1433.

[4] Song, B., J. Kim, and S. Oh, "Performance Analysis of Channel Assignment Methods for Multiple Carrier CDMA Cellular Systems," *IEEE Proc. of VTC*, Spring 1999, pp. 10–14.

[5] Kleinrock, L., *Queueing Systems, Vol. 1: Theory*, New York: John Wiley & Sons, 1975.

CHAPTER 11

Erlang Capacity of Hybrid FDMA/CDMA Systems Supporting Multiclass Services

Future mobile networks will consist of several distinct radio access technologies, such as WCDMA or GSM/EDGE, where each radio access technology is denoted as a *subsystem*. These future wireless networks, which demand the cooperative use of a multitude of subsystems, are named *multiaccess systems*. In multiservice scenarios, the overall capacity of multiaccess networks depends on how users of different types of services are assigned on to subsystems because each subsystem has its own distinct features in the aspect of capacity. For example, IS-95A can handle voice service more efficiently than data service, while WCDMA can handle data service more efficiently than voice service.

In this book, we tackle the Erlang capacity evaluation of multiaccess systems in the following two cases. First, in this chapter, we consider the case that each subsystem provides similar air link capacity as with hybrid FDMA/CDMA, where like FDMA, the available wideband spectrum of the hybrid FDMA/CDMA is divided into a number of distinct bands. Each connection is allocated to a single band such that each band facilitates a separate narrowband CDMA system whose signals employ DS spreading and are transmitted in one and only one band.

Typically each band has a bandwidth of 1.25 MHz for compatibility with IS-95A. For hybrid FDMA/CDMA, because the carriers are colocated, they all experience an identical topological and RF environment such that it can usually be assumed that each carrier will provide similar air link capacity.

Second, in Chapter 12, we will consider the case that each subsystem provides different air link capacity as with the case with coexisting GSM/EDGE-like and WCDMA-like subsystems. In this case, the overall capacity of multiaccess networks depends on the employed service assignment (i.e., the way of assigning users of different types of services on to subsystems).

To evaluate the Erlang capacity in the latter case is more complicated than in the first case because the service assignment scheme should be involved in the capacity analysis.

11.1 Introduction

CDMA has been widely studied in the past two decades due to its superior voice quality, robust performance, and large air interface capacity. Commercial CDMA systems have been already launched and operated successfully. A typical example of these commercial systems is IS-95. Existing IS-95-based CDMA systems support circuit mode and packet mode data services at a data rate limited to 9.6–14.4 Kbps. Many wireless data applications that do not need higher data rates will operate efficiently and economically by using these systems [1]. Future mobile communications systems, however, will provide not only voice and low-speed data services, but also video and high-speech data services. To support these multimedia services, higher capacity and higher data rates should be guaranteed. Hybrid FDMA/CDMA, proposed in [2], is currently being considered as a promising approach for third-generation mobile and personal communication systems. In hybrid FDMA/CDMA, like FDMA, the available wideband spectrum is divided into a number of distinct bands. Each connection is allocated to a single band such that each band facilitates a separate narrowband CDMA system, whose signals employ DS spreading and are transmitted in one and only one band. Typically each band has a bandwidth of 1.25 MHz for compatibility with the IS-95A. Also, distinct bands are carried by different carriers.

For hybrid FDMA/CDMA, because the carriers are co-located, they experience an identical topological and RF environment. It is usually assumed that each carrier will provide similar air link capacity. In this case, the performance of the hybrid FDMA/CDMA can be varied with the channel assignment methods. There are typically two channel assignment methods applicable to hybrid FDMA/CDMA cellular systems that support multiple carriers: ICCA and CCCA [3]. In ICCA, traffic channels of each carrier are handled independently so that each MS is allocated a traffic channel of the same carrier that it used in its idle state. By contrast, the CCCA scheme combines all traffic channels in the system. When a BS receives a new call request, a BS searches for the least occupied carrier and allocates a traffic channel in that carrier. Even through it is expected that the performance of hybrid FDMA/CDMA systems with CCCA schemes might be larger than that of hybrid FDMA/CDMA with ICCA due to the increased flexibility, there have been considerable interests in the quantitative performance comparison between ICCA and CCCA. In [3], Song et al. analyzed and compared performances of hybrid FDMA/CDMA systems under ICCA and CCCA schemes. However, they focused on the voice-oriented system. In addition, they considered the call blocking model in which the call blocking is caused only by a scarcity of the CEs that perform the baseband spread spectrum signal processing for the given channel in the BS. Practically, call blocking in hybrid FDMA/CDMA systems is caused not only by the scarcity of CEs in the BS but also by insufficient availability of channels per sector.

In this chapter, we present an analytical procedure to analyze the Erlang capacity for the hybrid FDMA/CDMA systems supporting voice and data services with multiple carriers of equal bandwidth under both ICCA and CCCA schemes. Here, we consider the expanded call blocking model in which call blocking is caused not only by the scarcity of CEs in the BS but also by insufficient available channels per sector. For the performance analysis, a multidimensional Markov chain model is developed, and the Erlang capacity is depicted as a function of the offered traffic loads of voice and data services. For each allocation scheme, the effect of the number of carriers of hybrid FDMA/CDMA systems on the Erlang capacity is observed, and the optimum values of the system parameters such as CEs are selected with respect to the Erlang capacity. Furthermore, the performances of ICCA are quantitatively compared with those of CCCA.

The remainder of this chapter is organized as follows: In Section 11.2, the system models are described. In Section 11.3, two channel assignment methods that can be applied to hybrid FDMA/CDMA systems are described. In Section 11.4, we present an analytical procedure to analyze the Erlang capacity of the hybrid FDMA/CDMA supporting voice and data services, based on the multidimensional Markov model. In Section 11.5, a numerical example is taken into consideration, and discussions are given. Finally, conclusions are drawn in Section 11.6.

11.2 System Model

For the performance analysis, the following assumptions are considered:

- We consider the hybrid FDMA/CDMA system supporting voice and data services and consisting of *P* carriers of equal bandwidth, where *P* denotes the number of the used carriers or bands in the system.
- The considered system employs directional antenna and divides a cell into a number of sectors to reduce multiuser interference. We consider a three-sector cell, by assuming perfect directional antennas. Further, all cells are equally loaded.
- At each sector, each carrier of hybrid FDMA/CDMA facilitates a narrowband CDMA system, whose signals employ DS spreading and are transmitted in one carrier. For each carrier facilitating a narrowband CDMA system, although there is no hard limit on the number of mobile users served, there is a practical limit on the number of concurrent users to control the interference between users that have the same pilot signal. The maximum number of concurrent users that a carrier can support with QoS requirements was found, based on the maximum tolerable interference [4, 5]. In particular, as a result of [5], the system capacity limit of a carrier in the reverse link can be expressed as:

$$\gamma_v N_v + \gamma_d N_d \leq 1 \tag{11.1}$$

where

$$\gamma_v = \frac{\alpha}{\frac{W}{R_{v_{req}}}\left(\frac{E_b}{N_o}\right)_{v_{req}}^{-1}\left\langle\frac{1}{1+f}\right\rangle 10^{\frac{Q^{-1}(\beta)}{10}\sigma_x - 0.012\sigma_x^2} + \alpha}$$

$$\gamma_d = \frac{1}{\frac{W}{R_{d_{req}}}\left(\frac{E_b}{N_o}\right)_{d_{req}}^{-1}\left\langle\frac{1}{1+f}\right\rangle 10^{\frac{Q^{-1}(\beta)}{10}\sigma_x - 0.012\sigma_x^2} + 1}$$

γ_v and γ_d are the amount of system resources that are used by one voice and one data user, respectively. N_v and N_d denote the number of users in the voice and data service groups, respectively; *W* is the allocated frequency bandwidth per carrier; σ_x is the standard deviation of the received SIR that indicates the overall effect of imperfect power control; $\beta\%$ is the system reliability; α is the voice activity factor; $(E_b/N_0)_{v_{req}}$ and $(E_b/N_0)_{d_{req}}$ are the required bit energy-to-interference power spectral density ratio for the voice and the data service groups, respectively; f is the other cell interference factor defined as the ratio of intercell interference from intercell to the intracell interference from intracell; $\left\langle \frac{1}{1+f} \right\rangle$ is the average value of frequency reuse factor; and Q^{-1} is the inverse Q function defined as $Q(x) = \int_{-\infty}^{x} \left(1/\sqrt{2\pi e}^{-y^2/2}\right) dy$.

Based on (11.1), it is assumed that each carrier of hybrid FDMA/CDMA provides $\hat{C}_{ETC}$ basic channels per sector, and the system resource that is used by one data call is equivalent to Λ times that of one voice call, where $\hat{C}_{ETC} \equiv \lfloor 1/\gamma_v \rfloor, \Lambda = \lfloor \gamma_d / \gamma_v \rfloor$ and $\lfloor x \rfloor$ denotes the greatest integer less than or equal to x.

- There are N CEs per cell, where all CEs are pooled in the BS such that any CE can be assigned to any user in the cell, regardless of sector. Basically, the CE performs the baseband spread spectrum signal processing for a given channel (pilot, sync, paging, or traffic channel) in the BS.
- The system employs a circuit switching method to deal with the traffic transmission for voice and data services. Each user shares the system resources with other users and competes with other users for the use of the system resources. In this situation, a call attempt may be blocked. We consider two types of call blocking models: hard blocking, which is caused by insufficient CEs in the BS, and soft blocking, which occurs when the number of active users exceeds the maximum number of basic channels in each sector. In addition, blocked calls are cleared.
- We assume that two call arrivals of voice and data traffics in the ith sector (i = 1, 2, 3) are distributed according to independent Poisson processes with average call arrival rate $\lambda_{(v,i)}$ and $\lambda_{(d,i)}$, respectively. Also, the channel holding times of voice and data traffic are exponentially distributed with mean channel holding time $1/\mu_{(v,i)}$ and $1/\mu_{(d,i)}$, respectively. Then, the traffic loads of voice and data services in the ith sector, $\rho_{(v,i)}$ and $\rho_{(d,i)}$, are given as $\lambda_{(v,i)}/\mu_{(v,i)}$ and $\lambda_{(d,i)}/\mu_{(d,i)}$, respectively.

11.3 Channel Assignment Methods

There are two main channel assignment methods for the hybrid FDMA/CDMA cellular systems: without carrier transition and with carrier transition [3]. In the no-carrier-transition method, when a BS receives a channel request from an MS of the mth carrier [i.e., an MS that uses the mth carrier in its idle state (m = 1, ..., P)], it allocates the MS a traffic channel of the mth carrier. On the contrary, in the second method, a BS may allocate a traffic channel in other carriers according to the traffic condition in each carrier.

11.3.1 ICCA

In the ICCA scheme, traffic channels in each carrier are handled independently, so that each MS is always allocated a traffic channel in the same carrier that it uses in its idle state as determined by the hash function. When a BS receives a channel request from an MS of the mth carrier, it allocates the MS a traffic channel in the mth carrier even in the case that the mth carrier is the most highly loaded one.

11.3.2 CCCA

The CCCA scheme combines all traffic channels in all carriers. When a BS receives a call request from an MS of the mth carrier ($m = 1, \ldots, P$), it searches the least loaded carrier and allocates a traffic channel in that carrier.

11.4 Erlang Capacity Analysis

In this section, we present an analytical procedure to analyze the Erlang capacity for the hybrid FDMA/CDMA systems with P carriers under both ICCA and CCCA schemes, based on the multidimensional $M/M/m$ loss model.

11.4.1 Erlang Capacity Analysis for CCCA

Each user shares the system resources with other users and competes with other users for the use of the system resources. In this situation, a call attempt may be blocked. We consider two types of call blocking model: hard blocking, which is caused by insufficient CEs in the BS, and soft blocking, which occurs when the number of active users exceeds the maximum number of basic channels in each sector. We denote the hard blocking probability of voice and data in the BS as $b_{(hard,\, v)}$ and $b_{(hard,\, d)}$, respectively, and the soft blocking probability of voice and data in the sector i as $b_{(soft,\, v,\, i)}$ and $b_{(soft,\, d,\, i)}$, respectively.

In the CCCA, arrival of call attempts in a carrier depends upon the status of other carriers' occupation, and all traffic channels in all carriers are combined. That is, in the CCCA scheme, a BS may allocate a traffic channel in other carriers according to the traffic condition in each carrier. In the overall aspect of the system, hybrid FDMA/CDMA systems with P carriers under the CCCA scheme conceptually support $\hat{C}_{ETC} \cdot P$ basic channels per sector if each carrier provides $\hat{C}_{ETC}$ basic channels. In this situation, in order for a call attempt to get the service, soft blocking of the call should not occur in each sector and the hard blocking of the call also should not occur in the BS.

In this chapter, we adopt the approximate analysis method proposed in Chapter 10. That is, we decouple the calculation stages of soft blocking and hard blocking for the simplicity of computation such that the soft blocking and hard blocking probabilities can be separable as a closed-form equation, respectively. Noting that these closed-form equations may not provide universal values of soft blocking and hard blocking probabilities because practically the soft blocking and hard blocking probabilities affect each other in the blocking model being considered, here we introduce the coupling parameters, $\bar{\rho}_{(v,i)}, \bar{\rho}_{(d,i)}, \alpha_v$, and α_d, to consider mutual effects between hard blocking in the BS and soft blocking in the ith sector.

First, let's consider the closed-form equation for the soft blocking probability in the ith sector, and let $\mathbf{N}_i = (n_{(v,i)}, n_{(d,i)})$be the state of the ith sector ($i = 1, 2, 3$), given by the number of calls of each service group in the ith sector.

Then, the state probability of $\mathbf{N}_i$ in the ith sector, given traffic loads of voice and data services, is given by [6]:

$$\pi_i(\mathrm{N_i}) = \begin{cases} \dfrac{1}{G_i(R)} \dfrac{\bar{\rho}_{(v,i)}^{\,n_{(v,i)}}}{n_{(v,i)}!} \dfrac{\bar{\rho}_{(d,i)}^{\,n_{(d,i)}}}{n_{(d,i)}!} & \mathrm{N_i} \in S_i(R) \\ 0 & \mathrm{N_i} \notin S_i(R) \end{cases} \tag{11.2}$$

where $\bar{\rho}_{(v,i)}$ and $\bar{\rho}_{(d,i)}$ are defined to consider the traffic load of voice and data in the ith sector, which are somewhat reduced from the given traffic load due to the limitation of CEs in the BS, respectively.

In (11.2), $G_i(R)$ is a normalizing constant for the ith sector state probability that has to be calculated in order to have $\pi_i(\mathbf{N}_i)$ that is accumulated to 1:

$$G_i(R) = \sum_{\mathrm{N_i} \in S_i(R)} \frac{\bar{\rho}_{(v,i)}^{\,n_{(v,i)}}}{n_{(v,i)}!} \frac{\bar{\rho}_{(d,i)}^{\,n_{(d,i)}}}{n_{(d,i)}!} \tag{11.3}$$

For a hybrid FDMA/CDMA system supporting voice and data services with P carriers under the CCCA scheme, a set of admissible states $S_i(R)$ in the ith sector is given as:

$$S_i(R) = \{\mathrm{N_i} \mid \mathrm{N_i}\, \mathbf{A}^T \le R\} \tag{11.4}$$

where $\mathbf{A}$ is a 1×2 vector whose elements are the amount of system resources used by one voice and one data user, respectively, and R is a scalar representing the sector resource such that

$$\mathbf{A} = [1 \;\; \Lambda] \tag{11.5}$$

$$R = \hat{C}_{ETC} \cdot P \tag{11.6}$$

Then, the soft blocking probabilities for voice and data services in the ith sector can be easily evaluated as following:

$$b_{(soft,v,i)} = 1 - \frac{G_i(R - \mathbf{A}e_v)}{G_i(R)} \tag{11.7}$$

$$b_{(soft,d,i)} = 1 - \frac{G_i(R - \mathbf{A}e_d)}{G_i(R)} \tag{11.8}$$

where $e_v = [1\ 0]^T$ and $e_d = [0\ 1]^T$. $G_i(R)$ is the normalizing constant calculated on the whole $S_i(R)$, while $G_i(R - \mathbf{A}e_v)$ and $G_i(R - \mathbf{A}e_d)$ are the constants calculated on the $S_i(R - \mathbf{A}e_v)$ and $S_i(R - \mathbf{A}e_d)$, respectively.

In order for the calls, which are not soft blocked in each sector, to get the services, there should be sufficient CEs in the BS to support those calls. If there are not sufficient CEs in the BS, those calls will be hard blocked.

Because all CEs available in the BS are pooled and assigned to any all call regardless of sectors, α_v and α_d are introduced to consider the traffic load of voice and data that are offered to the BS from each sector and further defined as (11.14) and (11.15), respectively. For the purpose of evaluating hard blocking probability in the BS, let $\mathbf{N}_b = (n_v, n_d)$ be the state of the BS, given by the number of voice and data calls in the BS. Then, the state probability of $\mathbf{N}_b$ in the BS is given by

$$\pi(\mathbf{N}_b) = \frac{1}{G_b(R_b)} \frac{\alpha_v^{n_v}}{n_v!} \frac{\alpha_d^{n_d}}{n_d!} \tag{11.9}$$

where $G_b(R_b)$ is a normalizing constant for the state probability of the BS that must be calculated in order to get $\pi(\mathbf{N}_b)$, which is accumulated to 1, and it is given as

$$G_b(R_b) = \sum_{\mathbf{N}_b \in S_b(R_b)} \frac{\alpha_v^{n_v}}{n_v!} \frac{\alpha_d^{n_d}}{n_d!} \tag{11.10}$$

$$S_b(R_b) = \{\mathbf{N}_b | \mathbf{N}_b \mathbf{A}^T \leq R_b\} \tag{11.11}$$

where $S_b(R_b)$ is a set of admissible states in the BS, $R_b = N$, and N is the total number of CEs available in the BS.

Then, when there are N CEs in the BS, similar to soft blocking case, the closed-form equations for the hard blocking probabilities of voice and data services in the BS are given as follows:

$$b_{(hard,v)} = 1 - \frac{G_b(R_b - \mathbf{A}e_v)}{G_b(R_b)} \tag{11.12}$$

$$b_{(hard,d)} = 1 - \frac{G_b(R_b - \mathbf{A}e_d)}{G_b(R_b)} \tag{11.13}$$

where $G_b(R_b)$ is the normalizing constant calculated on the whole $S_b(R_b)$, while $G_b(R_b - \mathbf{A}e_v)$ and $G_b(R_b - \mathbf{A}e_d)$ are the constants calculated on the $S_b(R_b - \mathbf{A}e_v)$ and $S_b(R_b - \mathbf{A}e_d)$, respectively.

Subsequently, to evaluate the soft blocking and hard blocking probabilities, we must solve (11.7) and (11.12) for voice, and (11.8) and (11.13) for data, respectively, which are mutually linked by coupling parameters $\bar{\rho}_{(v,i)}$, $\bar{\rho}_{(d,i)}$, α_v and α_d. For the calculation of these blocking probabilities, in this chapter, we adopt an iteration method, which is described in Figure 11.1. Here, we let $b_{(soft,\,v,\,i)}(m)$, $b_{(soft,\,d,\,i)}(m)$, $b_{(hard,\,v)}(m)$, and $b_{(hard,\,d)}(m)$ represent the value of $b_{(soft,\,v,\,i)}$, $b_{(soft,\,d,\,i)}$, $b_{(hard,\,v)}$, and $b_{(hard,\,d)}$ at the mth iteration, respectively, and let $b_{(soft,\,v,\,i)}(0)$, $b_{(soft,\,d,\,i)}(0)$, $b_{(hard,\,v)}(0)$, and $b_{(hard,\,d)}(0)$ be the

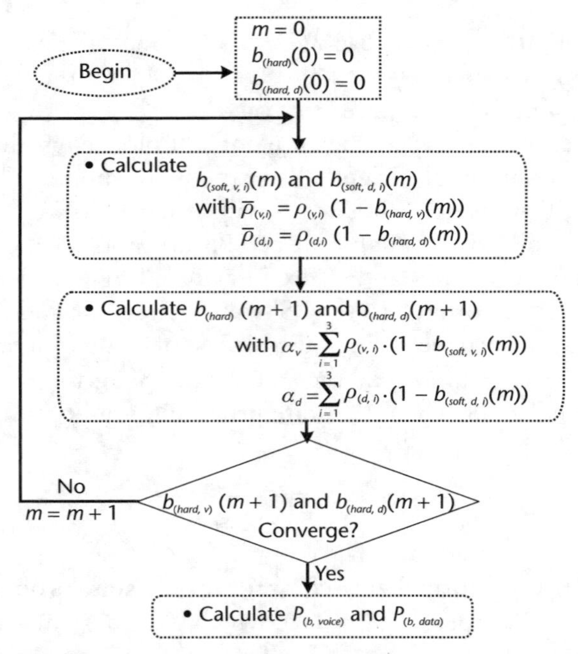

Figure 11.1 Iteration method to compute the blocking probability.

initial value for the recursion. At the mth iteration, $b_{(soft,\,v,\,i)}(m)$ and $b_{(soft,\,d,\,i)}(m)$ are computed using (11.7) and (11.8) with $\overline{\rho}_{(v,i)} = \rho_{(v,i)} \cdot \left(1 - b_{(hard,v)}(m)\right)$ and $\overline{\rho}_{(d,i)} = \rho_{(d,i)} \cdot \left(1 - b_{(hard,d)}(m)\right)$, respectively, where we intuitively let $\overline{\rho}_{(v,i)}$ and $\overline{\rho}_{(d,i)}$ as $\rho_{(v,i)} \cdot \left(1 - b_{(hard,v)}(m)\right)$ and $\rho_{(d,i)} \cdot \left(1 - b_{(hard,d)}(m)\right)$ reflect on the effect of the limited number of CEs (N) in the BS on the soft blocking probability in the ith sector through the feedback quantity of $b_{(hard,\,v)}(m)$ and $b_{(hard,\,d)}(m)$. Also, at the mth iteration, $b_{(hard,\,v)}(m)$ and $b_{(hard,\,d)}(m)$ are computed using (11.12) and (11.13) with the following α_v and α_d.

$$\alpha_v = \sum_{i=1}^{3} \rho_{(v,i)} \cdot \left(1 - b_{(soft,v,i)}(m-1)\right) \tag{11.14}$$

$$\alpha_d = \sum_{i=1}^{3} \rho_{(d,i)} \cdot \left(1 - b_{(soft,d,i)}(m-1)\right) \tag{11.15}$$

where we also intuitively let α_v and α_d be $\sum_{i=1}^{3} \rho_{(v,i)} \cdot \left(1 - b_{(soft,v,i)}(m)\right)$ and $\sum_{i=1}^{3} \rho_{(d,i)} \cdot \left(1 - b_{(soft,d,i)}(m)\right)$, respectively, to consider the effect of the user limit in each sector and the traffic loads of each sector on the hard blocking probability in the BS. Then, the iteration procedure takes the following steps.

1. Define $m = 0$ and set $b_{(hard,\,v)}(0) = 0$ and $b_{(hard,\,d)}(0) = 0$.

2. Calculate $b_{(soft,v,i)}(m)$ and $b_{(soft,d,i)}(m)$ for all i (i = 1, 2, 3) with $\bar{\rho}_{(v,i)} = \rho_{(v,i)} \cdot \left(1 - b_{(hard,v)}(m)\right)$ and $\bar{\rho}_{(d,i)} = \rho_{(d,i)} \cdot \left(1 - b_{(hard,d)}(m)\right)$.
3. Calculate $b_{(hard,v)}(m+1)$ and $b_{(hard,d)}(m+1)$ with $\alpha_v = \sum_{i=1}^{3} \rho_{(v,i)} \cdot \left(1 - b_{(soft,v,i)}(m)\right)$ and $\alpha_d = \sum_{i=1}^{3} \rho_{(d,i)} \cdot \left(1 - b_{(soft,d,i)}(m)\right)$.
4. If $\left(\left|b_{(hard,v)}(m+1) - b_{(hard,v)}(m)\right| / b_{(hard,v)}(m+1)\right) < \tau$ (tolerance parameter) and $\left(\left|b_{(hard,d)}(m+1) - b_{(hard,d)}(m)\right| / b_{(hard,d)}(m+1)\right) < \tau$, then stop the recursion. Otherwise, set $m = m + 1$ and go back to step 2.

From our numerical experiences, it is observed that this recursion always converges within a few iterations (generally less than five). Also, it is noteworthy that even though we select the coupling parameters $\bar{\rho}_{(v,i)}$, $\bar{\rho}_{(d,i)}$, α_v, and α_d somewhat intuitively, the other forms of coupling parameters may be adopted for the better calculation of soft blocking and hard blocking probabilities.

Finally, the call blocking probabilities of voice and data services in the ith sector, $P_{(b,\ voice)}$ and $P_{(b,\ data)}$ are given as follows for convergence values.

$$\begin{aligned} P_{(b,voice)} &= 1 - \left(1 - b_{(soft,v,i)}\right) \cdot \left(1 - b_{(hard,v)}\right) \\ &= 1 - \frac{G_i(R - \mathbf{A}e_v)}{G_i(R)} \cdot \frac{G_b(R_b - \mathbf{A}e_v)}{G_b(R_b)} \end{aligned} \tag{11.16}$$

$$\begin{aligned} P_{(b,data)} &= 1 - \left(1 - b_{(soft,d,i)}\right) \cdot \left(1 - b_{(hard,d)}\right) \\ &= 1 - \frac{G_i(R - \mathbf{A}e_d)}{G_i(R)} \cdot \frac{G_b(R_b - \mathbf{A}e_d)}{G_b(R_b)} \end{aligned} \tag{11.17}$$

In the hybrid FDMA/CDMA systems supporting voice and data services, Erlang capacity corresponding to the voice-only system needs to be modified in a vector format to consider the performance of two distinct service groups simultaneously. In this chapter, Erlang capacity is defined as a set of the average offered traffic loads of each service group that can be supported while the QoS and GoS requirements are being satisfied. Then, Erlang capacity at the ith sector, C_{Erlang}, can be calculated as following:

$$\begin{aligned} C_{Erlang} &= \left\{\left(\hat{\rho}_{(v,i)}, \hat{\rho}_{(d,i)}\right)\right\} \\ &= \left\{\left(\rho_{(v,i)}, \rho_{(d,i)}\right) \mid P_{(b,voice)} \le P_{(B,v)_{req}}, P_{(b,data)} \le P_{(B,d)_{req}}\right\} \end{aligned} \tag{11.18}$$

$P_{(B,\ v)_{req}}$ and $P_{(B,\ d)_{req}}$ are the required call blocking probability of voice and data calls, respectively, which can be considered GoS requirements.

In other words, the system Erlang capacity in the *i*th sector is a set of values of $\{(\hat{\rho}_{(v,i)}, \hat{\rho}_{(d,i)})\}$ that keep the call blocking probability experienced by each call less than the required call blocking probability (or GoS requirements) of each call.

11.4.2 Erlang Capacity Analysis for ICCA

In the case of ICCA, the Erlang capacity of an arbitrary carrier represents the performance of a hybrid FDMA/CDMA system because each carrier operates independently.

Subsequently, the Erlang capacity of hybrid FDMA/CDMA with *P* carriers is the product of *P* and the Erlang capacity of an arbitrary carrier. Here, let's denote $C_{Erlang/1FA}$ as Erlang capacity of an arbitrary carrier in the ICCA scheme.

$C_{Erlang/1FA}$ can be calculated by replacing (11.6), (11.14), and (11.15) with (11.19), (11.20), and (11.21), respectively, and then repeating the procedures applied in Section 11.4.1.

$$R = \hat{C}_{ETC} \tag{11.19}$$

$$\alpha_v = \sum_{i=1}^{3} \rho_{(v,i)} \cdot \left(1 - b_{(soft,v,i)}(m)\right) \cdot P \tag{11.20}$$

$$\alpha_d = \sum_{i=1}^{3} \rho_{(d,i)} \cdot \left(1 - b_{(soft,d,i)}(m)\right) \cdot P \tag{11.21}$$

Finally, the Erlang capacity of hybrid FDMA/CDMA with *P* carriers under ICCA is given as $P \cdot C_{Erlang/1FA}$.

11.5 Numerical Example

As a numerical example, we consider a hybrid FDMA/CDMA system supporting voice and data services with *P* carriers of 1.25 MHz in the three-sector cells ($P = 1, 2, 3, 4$, or 5). The system parameters under the consideration are given in Table 11.1.

In this example, each carrier can individually provides 29 basic channels per sector, based on (11.1), and the system resource used by one data call is equivalent to six times that of one voice call, such that $\hat{C}_{ETC}$ and Λ are given as 29 and 6, respectively.

Figure 11.2 shows the Erlang capacities of ICCA and CCCA for different values of CEs, respectively, when $P = 2$, and $P_{(B,v)_{req}}$ and $P_{(B,d)_{req}}$ are all given as 2%. The dotted lines are the Erlang capacities of CCCA and the solid lines are those of ICCA. All points $(\hat{\rho}_v, \hat{\rho}_d)$ under each Erlang capacity line represent the supportable offered traffic loads of voice and data services while QoS and GoS requirements are being satisfied. Figure 11.2 shows that the Erlang capacities between CCCA and ICCA are almost same when the number of CEs is small. The reason is that the flexibility of CCCA, which comes from the combination of all traffic channels in all carriers at each sector, has no influence on Erlang capacity for the small CEs because the call

Table 11.1 System Parameters for the Hybrid FDMA/CDMA System Supporting Voice and Data Services

Parameters	*Symbol*	*Value*
Allocated frequency bandwidth	W	1.25 Mbps
Number of the carriers of hybrid FDMA/CDMA	P	Variable
Required bit transmission rate for voice traffic	R_v	9.6 Kbps
Required bit transmission rate for data traffic	R_d	28.8 Kbps
Required bit energy-to-interference power spectral density ratio for voice traffic	$\left(\frac{E_b}{N_o}\right)_{v_{req}}$	7 dB
Required bit energy-to-interference power spectral density ratio for data traffic	$\left(\frac{E_b}{N_o}\right)_{d_{req}}$	7 dB
System reliability	$\beta\%$	99%
Frequency reuse factor	$\left\langle\frac{1}{1+f}\right\rangle$	0.7
Standard deviation of received SIR	σ_x	1dB
Voice activity factor	α	3/8

blocking of voice and data calls mainly occurs due to insufficient CEs in the BS. However, as the number of CEs available in the BS increases, call blocking of voice and data calls gradually occurs due not to insufficient CEs in the BS but to user limit per sector. Subsequently, CCCA improves the call blocking probabilities of voice and data calls by pooling the capacity offered by the individual carrier per sector and further outperforms ICCA for a larger number of CEs. This fact can be observed in Figure 11.2.

It is intuitive that the more CEs there are, the larger Erlang capacity will be. However, the Erlang capacity will be saturated after a certain value of CEs due to insufficient traffic channels per sector. For deeper consideration of the effect of CEs on Erlang capacity, we assume that the offered traffic load of data is proportional to

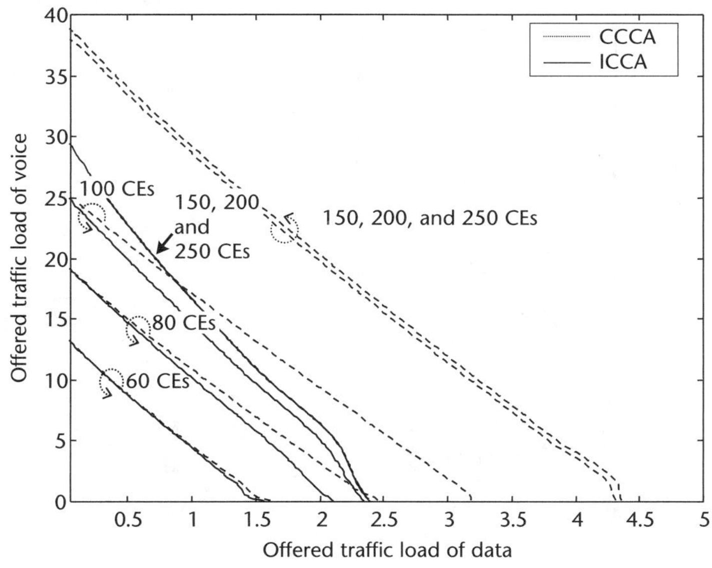

Figure 11.2 Erlang capacities of CCCA and ICCA for different values of CEs when the number of carriers of hybrid FDMA/CDMA, P, is 2.

that of voice and let δ be the traffic ratio of data to voice by which the dimension of Erlang capacity can be reduced into one dimension. Figure 11.3 shows Erlang capacity as a function of the number of CEs when $P = 2$ and $\delta = 1\%$. From Figure 11.3, we observe that the Erlang capacity region can be divided into three regions.

In the first region, Erlang capacity increases linearly with the increase of the CEs. This means that call blocking, in this region, occurs mainly due to the limitation of CEs in the BS. In the second region, Erlang capacity is determined by the interplay between the limitation of CEs in the BS and insufficient traffic channels per sector. Finally, in the last region, Erlang capacity is saturated where call blocking is mainly due to insufficient traffic channels per sector. Figure 11.3 also shows that Erlang capacity of ICCA is more quickly saturated than that of CCCA.

Figure 11.4 shows Erlang capacity according to the number of carriers of hybrid FDMA/CDMA system when $\delta = 1\%$. As the number of carriers of hybrid FDMA/CDMA system increases, the maximum achievable Erlang capacities for both ICCA and CCCA schemes are also increased, respectively. Generally, it is an interesting question to the system operator to estimate the number of carriers that are required to accommodate the target traffic loads. For example, if there are the voice traffic loads of 50 Erlang and data traffic loads of 1 Erlang per sector, respectively, which corresponds to $\delta = 1\%$, there might be a question of how many carriers of hybrid FDMA/CDMA are needed to support these traffic loads. To this question, we recommend using at least three carriers for CCCA schemes and at least four carriers for ICCA, based on Figure 11.4. Also, Figure 11.4 shows CCCA outperforms ICCA with the increase of carriers.

Figure 11.5 shows Erlang capacity increments of CCCA over ICCA as a function of the number of CEs for the different numbers of carriers. For each number of

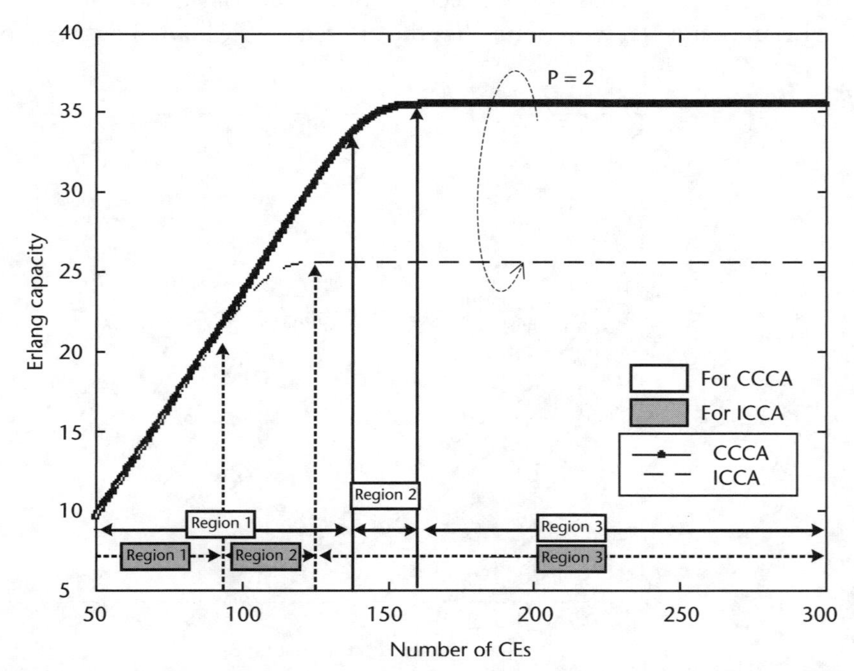

Figure 11.3 Erlang capacity as a function of the number of CEs when the number of carriers of hybrid FDMA/CDMA, P, is two and the traffic ratio of data to voice, δ, is 1%.

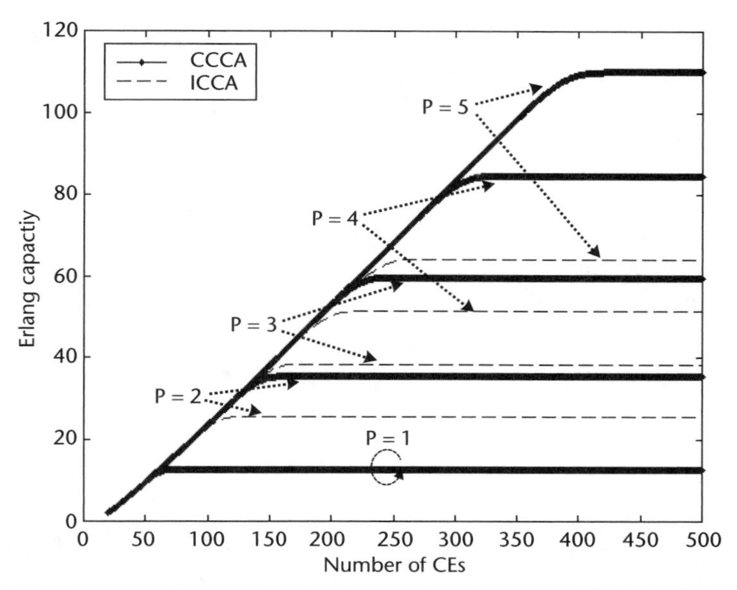

Figure 11.4 Erlang capacity according to the number of carriers of hybrid FDMA /CDMA systems when the traffic ratio of data to voice, δ, is 1%.

carriers, the Erlang capacity of CCCA is almost same as that of ICCA with same CEs, while CCCA outperforms ICCA with an increase of CEs. Finally, if there are enough CEs in the BS, Erlang capacity is maximally improved by 38%, 55%, 64%, and 74% using CCCA when the number of multiple carriers, P, is 2, 3, 4, and 5, respectively. However, it is noted that even though CCCA shows a higher Erlang capacity than ICCA, it requires more control information, such as the carrier's channel occupation status.

Another important performance measure is =CE utilization. This is defined as C_{Erlang}/N, where C_{Erlang} denotes the Erlang capacity of the hybrid FDMA/CDMA system, and N is the number of CE available in the BS.

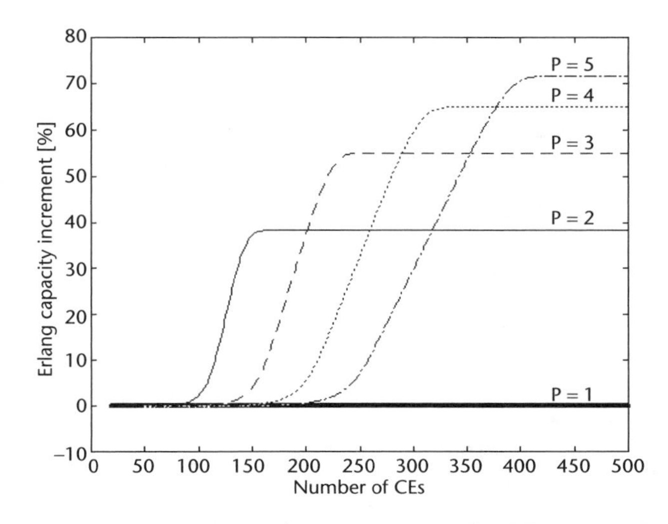

Figure 11.5 Erlang capacity increments of CCCA over ICCA for different numbers of carriers.

By definition, the CE utilization measures the average number of subscribers that each CE in each cell can accommodate. Figure 11.6 shows the CE utilization of CCCA and ICCA when $\delta = 1\%$. The solid line indicates the CE utilization of CCCA, and the dotted line indicates that of ICCA. Figure 11.6 shows that CEs are more efficiently used in CCCA as a consequence of the capacity improvement. For the given carriers and the considered channel assignment schemes, we can also find the optimum value of CE, N_{opt}, with respect to CE utilization. Table 11.2 shows N_{opt} and the corresponding CE utilization of ICCA and CCCA, respectively. Practically, the hybrid FDMA/CDMA is equipped with a finite number of CEs, offered by the "cost-efficient" system strategy. It is interesting to the system operators to select the optimum value of CEs with which CE utilization is maximized.

11.6 Conclusion

In this chapter, we present an analytical procedure for evaluating the Erlang capacity of hybrid FDMA/CDMA systems supporting voice and data services under two channel assignment methods: ICCA and CCCA. For each allocation method, the Erlang capacity of a hybrid FDMA/CDMA system is depicted as a function of the offered traffic loads of voice and data. The CCCA scheme shows considerable Erlang capacity improvement with the increase of carriers of a hybrid FDMA/CDMA system. For a fixed number of carriers, the Erlang capacity of CCCA is almost same as that of ICCA when the number of CEs is small. However, CCCA outperforms ICCA as the number of CEs increases. In the case of the numerical example, it is observed that the Erlang capacity is maximally improved by 38%, 55%, 65%, and 74% using CCCA when the traffic ratio of data to voice, δ, is 1% and the number of multiple carriers, P, is 2, 3, 4, and 5, respectively. Finally, we expect that the results of this chapter can be utilized for the traffic engineer to determine the required number of CDMA carriers in each sector and the required number

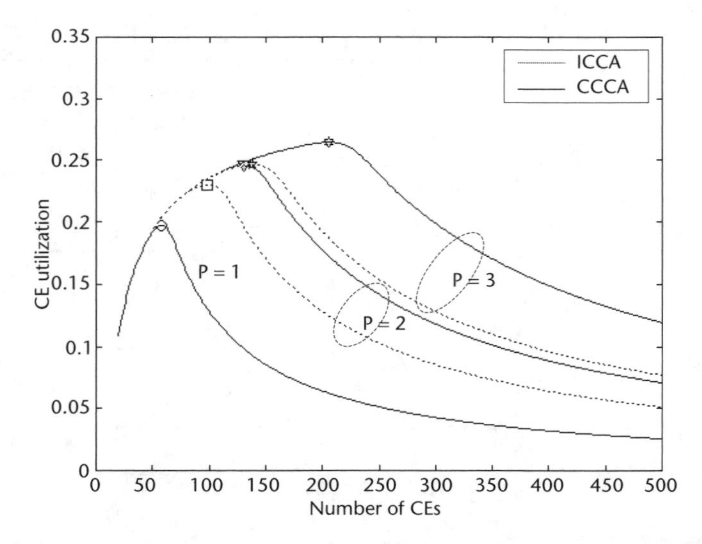

Figure 11.6 CE utilization of CCCA and ICCA when the traffic ratio of data to voice, δ, is 1%.

Table 11.2 Optimum Values of CEs with Respect to CE Utilization and the Corresponding CE Utilization

	ICCA		*CCCA*		
P	N_{opt}	*CE Utilization at* N_{opt}	N_{opt}	*CE Utilization at* N_{opt}	*CE Utilization Increment at* N_{opt}
1	58	0.1974	58	0.1974	0%
2	98	0.2304	130	0.2458	6.7%
3	138	0.2467	206	0.2646	7.2%

of CEs in the BS in order to accommodate the target traffic loads for each allocation method.

References

[1] Knisely, D. N., et al. "Evolution of Wireless Data Services: IS-95 to CDMA2000," *IEEE Communications Magazine*, 1998, pp. 140–149.

[2] Eng, T., and L. B. Milstein, "Comparison of Hybrid FDMA/CDMA Systems in Frequency Selective Rayleigh Fading," *IEEE Journal of Selected Areas in Communications*, 1994, pp. 938–951.

[3] Song, B., J. Kim, and S. Oh, "Performance Analysis of Channel Assignment Methods for Multiple Carrier CDMA Cellular Systems," *IEEE Proc. of VTC*, Spring 1999, pp. 10–14.

[4] Yang, Y. R., et al., "Capacity Plane of CDMA System for Multimedia Traffic," *IEEE Electronics Letters*, 1997, pp. 1432–1433.

[5] Koo, I., et al., "A Generalized Capacity Formula for the Multimedia DS-CDMA System," *IEEE Proc. of Asia-Pacific Conference on Communications*, 1997, pp. 46–50.

[6] Kelly, F., "Loss Networks," *The Annals of Applied Probability*, 1991, pp. 319–378.

CHAPTER 12

Erlang Capacity of Multiaccess Systems Supporting Voice and Data Services

In this chapter, we analyze and compare the Erlang capacity of multiaccess systems supporting several different radio access technologies according to two different operation methods—separate and common operation methods—by simultaneously considering the link capacity limit per sector as well as CE limit in BS. In a numerical example with GSM/EDGE-like and WCDMA-like subsystems, it is shown that we can get up to 60% Erlang capacity improvement through the common operation method when using a near optimum so-called service-based user assignment scheme, and there is no CE limit in BS. Even with the worst-case assignment scheme, we can still get about 15% capacity improvement over the separate operation method. However, the limited number of CEs in the BS reduces the capacity gains of multiaccess systems with the common operation over the separate operation. In order to fully extract the Erlang capacity of multiaccess system, an efficient method is needed to select the proper number of CEs in the BS while minimizing the cost of equipment.

12.1 Introduction

Future mobile networks will consist of several distinct radio access technologies, such as WCDMA or GSM/EDGE, where each radio access technology is denoted as a subsystem. Such future wireless networks, which demand the utilization of the cooperative use of a multitude of subsystems, are named multiaccess systems. In the first phase of such multiaccess systems, the RRM of subsystems may be performed in a separate way to improve the performance of individual systems independently, mainly because the subsystems have no information of the situation in other subsystems and the terminals do not have multimode capabilities. Under such a separate operation method, an access attempt is only accepted by its designated subsystem if possible; otherwise, it is rejected.

Intuitively, improvement of multiple-access systems is expected in a form of common resource management, where the transceiver equipment of the mobile stations supports multimode functions such that any terminal can connect to any subsystem. This may be accomplished either through parallel transceivers in hardware or by using software radio [1]. The common RRM functions may be implemented in existing system nodes, but interradio access technology signaling mechanisms need to be introduced. In order to estimate the benefit of such common resource management of multiaccess systems, some studies are necessary, especially with regard to quantifying the associated Erlang capacity.

As an example of improving the performances of common resource management for single-service scenarios, the *trunking gain* of multiaccess system capacity enabled by the larger resource pool of common resource management has previously been evaluated in [2] by relatively simple Matlab-based simulations, and multiservice allocation is not considered. In multiservice scenarios, it is expected that the capacity of multiaccess systems also depends on how users of different types of services are assigned on to subsystems. The gain that can be obtained through the employed assignment scheme can be named the *assignment gain*, and further the capacity gain achievable with different user assignment principles has been estimated in [3–5]. These studies, however, disregard trunking gains.

In this section, we combine these two approaches to analysis and further quantify the capacity gain of multiaccess systems by simultaneously considering the trunking gain and the assignment gain. More specifically, we focus on analyzing and comparing the Erlang capacity of multiaccess systems supporting voice and data services according to two operation methods: separate and common operation methods. In the case of the common operation method, we also consider two user assignment schemes: the service-based assignment algorithm [3] as a best case reference, which roughly speaking assigns users to the subsystem where their service is most efficiently handled, and the rule opposite to the service-based assignment as a worst case reference.

When analyzing the Erlang capacity of mulitaccess system, we also consider two resource limits simultaneously—link capacity limit per sector and CE limit in the BS—because practically a call blocking is caused by these two factors. However, most studies [2, 4, 5] do not consider the hardware limit at the BS, such as CEs, but mainly take into account the link capacity when evaluating the Erlang capacity. The issue of determining the proper number of CEs in a BS is critical to operators who wish to operate the system more cost efficiently because CEs are a cost part of the system. In the aspect, this chapter can provide a good guideline for operating and dimensioning the multiaccess systems.

The remainder of this chapter is organized as follows. In Section 12.2, the system model is described. In Section 12.3, two operation methods of the multiaccess system under consideration are described. In Section 12.4, we present an analytical procedure for analyzing the Erlang capacity of multiaccess systems according to the two operation methods. In Section 12.5, a numerical example is taken into consideration, and discussions are given. Finally, conclusions are drawn in Section 12.6.

12.2 System Model

For the performance analysis, following system model is considered:

- We consider the multiaccess system supporting voice and data services and consisting of P subsystems, where P denotes the number of the subsystems, and each subsystem provides its own link capacity. Each user is classified by QoS requirements such as the required transmission rate and BER, and all users in the same service group have the same QoS requirements.

- We consider the multiaccess system supporting voice and data services and consisting of P subsystems, where P denotes the number of the subsystems, and each subsystem provides its own link capacity. These user groups are classified by QoS requirements, such as different transmission rates and quality (BER) requirements.
- The considered system employs directional antenna and divides a cell into a number of sectors to reduce the multiuser interference. We consider a three-sector cell with perfect directional antennas and assume all cells are equally loaded.
- In the aspect of network operation, it is of vital importance to set up a suitable policy for the acceptance of an incoming call in order to guarantee a certain QoS. In general, CAC policies can be divided into two categories: NCAC and ICAC [6]. In the case of ICAC, a BS determines whether a new call is acceptable by monitoring the interference level on a call-by-call basis, while the NCAC utilizes a predetermined CAC threshold. In this section, we adopt a NCAC-type CAC based on its simplicity with which we can apply a general loss network model to the system being considered for the performance analysis, even though the NCAC generally suffers a slight performance degradation over the ICAC [6].
- Two resource limitations are also considered: the CE limitation in BSs and link capacity limitation per sector. The CE in the BS, an important hardware element, performs the baseband signal processing for a given channel in the BS. On the other hand, the link capacity limitation per sector is like a capacity with respect to the number of supportable current users. These limitations eventually result in call blocking, and here we consider two types of call blocking models: hard blocking, defined as call blocking that occurs when all CEs in the BS are used, and link blocking, defined as call blocking that occurs when the number of active users is equal to or exceeds the maximum number of basic channels in a particular sector. In particular, link blocking corresponds to soft blocking when the system under consideration is a CDMA-based system. We also denote the hard blocking probability of the call in the jth service group as $b_{(hard,\, j)}$ and the link blocking probability of the call in the jth service group in the sector i as $b_{(link,\, j,\, i)}$.
- For a constraint on the number of CE, we consider N CEs per cell or BS, where N denotes the total number of CEs available in the BS. The CE is a hardware element that performs the baseband signal processing for a given channel in the BS. Here it is noteworthy that CEs in the BS are a crucial cost part of the system such that they should be pooled in BS, and any CE can be assigned to any call in the cell regardless of its sector.
- In order to consider the link capacity limitation of multiaccess systems per sector, first we need to identify the admissible region of voice and data service groups in each subsystem. Let Q_v^l and Q_d^l be the link qualities, such as frame error rate, that individual voice and data users experience in the subsystem

l ($l = 1, 2, ..., P$), respectively, and $Q_{v,\min}$ and $Q_{d,\min}$ be a set of minimum link quality level of each service. Then, for a certain set of system parameters, such as service quality requirements, link propagation model, and system assumption, the admissible region of the subsystem l with respect to the simultaneous number of users satisfying service quality requirements in the sense of statistic $S_{sub,l}$ can be defined as

$$
\begin{aligned}
S_{sub,l} & \\
&= \left\{ \left(n_{(v,l)}, n_{(d,l)}\right) \mid P_r\left(Q_v^l \geq Q_{v,\min} \text{ and } Q_d^l \geq Q_{d,\min}\right) \geq \beta\% \right\} \\
&= \left\{ \left(n_{(v,l)}, n_{(d,l)}\right) \mid 0 \leq f_l\left(n_{(v,l)}, n_{(d,l)}\right) \leq 1 \text{ and } n_{(v,l)}, n_{(d,l)} \in Z_+ \right\} \text{ for } l = 1, 2, \ldots, P
\end{aligned}
\tag{12.1}
$$

where $n_{(v,l)}$ and $n_{(d,l)}$ are the admissible number of calls of voice and data service groups in the subsystem l, respectively; $\beta\%$ is system reliability defined as a minimum requirement on the probability that the link quality of the current users in the subsystem l is larger than the minimum link quality level, which is usually given between 95% and 99%; and $f_l(n_{(v,l)}, n_{(d,l)})$ is the normalized capacity equation of the subsystem l. In the case of a linear capacity region, for example, $f_l(n_{(v,l)}, n_{(d,l)})$ can be given as $f_l(n_{(v,l)}, n_{(d,l)}) = a_{lv} \cdot n_{(v,l)} + a_{ld} \cdot n_{(d,l)}$ for $l = 1, 2$. Such linear bounds on the total number of users of each class that can be supported simultaneously while maintaining adequate QoS requirements are commonly found in the other literature for CDMA systems supporting multiclass services [7, 8]. Further, provided the network state lines within the admissible region, then the QoS requirement of each user will be satisfied with $\beta\%$ reliability. When the admissible region of voice and data service groups in each subsystem is identified, the admission region of multiaccess systems varies according to the operation methods, on which more details will be given in Section 12.4.

In order to focus on the traffic analysis of subsystems under the CAC policy of our interest, we also consider the standard assumptions on the user arrival and departure processes. That is, we assume that call arrivals from users of class j in the subsystem l are generated as a Poisson process with rate $\lambda_{(j,l)}$ ($j = v, d$).

If a call is accepted, then it remains in the cell and subsystem of its origin for an exponentially distributed holding time with mean $1/\mu_{(j,l)}$, which is independent of other holding times and of the arrival processes. Then, the offered traffic load of the jth service group in the subsystem l is defined as $\rho_{(j,l)} = \lambda_{(j,l)}/\mu_{(j,l)}$.

12.3 Operation Methods of Multiaccess Systems

The overall performance of multiaccess system will depend highly on the operation methods. However, the operation of multiaccess systems will be limited by such factors as the terminal and network capabilities of supporting multimode function.

Here, we consider two extreme cases. One is the case that all terminals cannot support the multimode function, and the other is that all terminals can support it, which corresponds to the separate and common operation methods of multiaccess systems, respectively. It is expected that these two extreme cases will provide the lower and upper bound of the Erlang capacity of multiaccess systems.

12.3.1 Separate Operation Method

In the separate operation method, subsystems in a multiaccess system are operated independently, mainly because all terminals do not support multimode operation. Subsequently, traffic channels in each subsystem are handled independently so that each terminal is always allocated a traffic channel in its designated subsystem. Somewhat simply, in the separate operation method of the multiaccess systems, an access attempt is accepted by its designated subsystem if possible and otherwise rejected.

12.3.2 Common Operation Method

In the common operation method, any terminal that has multimode function can connect to any subsystem, such that air link capacities in all subsystems can be pooled, as with the case of the CCCA scheme of the hybrid FDMA/CDMA. However, the difference is that each subsystem provides a different air link capacity, as with the case with coexisting GSM/EDGE-like and WCDMA-like subsystems. In this case, the overall capacity of multiaccess networks depends on the employed service assignment (i.e., the way that users of different types of services are assigned onto subsystems).

In this chapter, we consider two user assignment schemes: the service-based assignment algorithm [3] as a best case reference, which roughly speaking assigns users to the subsystem where their service is most efficiently handled, and the rule opposite to the service-based assignment as a worst case reference.

- *The service-based assignment.* In [3], Furuskar discussed principles for allocating multiple services onto different subsystems in multiaccess wireless systems and further derived the favorable optimum subsystem service allocation scheme through simple optimization procedures that maximizes the combined capacity, which here is named *service-based assignment algorithm*. In the service-based assignment algorithm, we assign users into the subsystem where their expected relative resource cost for the bearer service type in question is the smallest.
- *The rule opposite to the service-based user assignment.* As the worst case in common operation, we consider the rule opposite to the service-based assignment scheme with which we assign users into the subsystem, where their expected relative resource cost for the bearer service type in question is the largest. Even though the rule opposite to service-based assignment is not likely to be used in reality, here we adopt it as an interesting reference for the worst case scenario of common operation.

These two extreme cases for user assignment will provide the upper and lower bounds of Erlang capacity under the common operations.

12.4 Erlang Capacity Analysis

In this section, we present a procedure for analyzing the Erlang capacity of multiaccess systems supporting voice and data services according to two different operation methods—separate and common operations—by simultaneously considering the link capacity limit per sector as well as the CE limit in the BS. The expressions are not in closed form but lend themselves to simple numerical methods using a few iterations. From the blocking probabilities, we could easily derive the Erlang capacity formulas. Let's first consider the case of the separate operation.

12.4.1 Erlang Capacity Analysis for Separate Operation Method

Due to the hardware limitation in BSs as well as link capacity limitation per sector, in order for a call attempt to get service, the link blocking of the call should not occur in a sector and the hard blocking of the call also should not occur in the BS. Here, we adopt the approximate analysis method proposed in [9] to evaluate the call blocking probability. That is, we decouple the calculation stages of link blocking and hard blocking for simplicity of computation such that the link blocking and hard blocking probabilities can be separable as closed-form equations. Noting that these closed-form equations may not provide universal values of link blocking and hard blocking probabilities because the link blocking and hard blocking probabilities practically affect each other in the blocking model being considered, here we introduce the coupling parameters $\bar{\rho}_{(j,l,i)}$ and α_i to consider mutual effects between the hard blocking in the BS and the link blocking in the ith sector.

Keeping in the mind that in the case of the separate operation method of the multiaccess systems, an access attempt is accepted by its designated subsystem if possible and otherwise rejected, let's first consider the close form for the link blocking probability of the lth subsystem in the ith sector and let $N_l^i \left(\equiv \left(n_{(v,l,i)}, n_{(d,l,i)}\right)\right)$ be state of the lth subsystem in the ith sector. With the system models and assumptions given in the previous sections, it is well known from *M/M/m* queue analysis that for given traffic loads, the equilibrium probability for an admissible state N_l^i in the subsystem $l, \pi(N_l^i)$ can have a product form on the truncated state space defined by the call admission strategy such that it is given by [10] (see Appendix B):

$$\pi(N_l^i) = \begin{cases} \dfrac{\dfrac{\bar{\rho}_{(v,l,i)}^{\,n_{(v,l,i)}} \bar{\rho}_{(d,l,i)}^{\,n_{(d,l,i)}}}{n_{(v,l,i)}!\, n_{(d,l,i)}!}}{\sum_{N_l^i \in S_{sub,l}} \dfrac{\bar{\rho}_{(v,l,i)}^{\,n_{(v,l,i)}} \bar{\rho}_{(d,l,i)}^{\,n_{(d,l,i)}}}{n_{(v,l,i)}!\, n_{(d,l,i)}!}} & N_l^i \in S_{sub,l} \\ 0 & \text{otherwise} \end{cases} \tag{12.2}$$

where $\bar{\rho}_{(j,l,i)}$ is introduced so as to consider the traffic load of the j service groups in the lth subsystem at the ith sector ($j = v,d$, $l = 1, ..., P$ and $i = 1,2,3$), which is somewhat reduced from the given traffic load due to the limitation of CEs in the BS. Then, the link blocking probability of a user of class j of the subsystem l at the sector i can simply expressed as

$$b_{(soft,j,l,i)} = \sum_{N_l^i \in S_{blk,l}^i} \pi(N_l^i) \tag{12.3}$$

where $S_{blk,l}^j$ is the subset of states in $S_{sub,l}$, whose states must move out of $S_{sub,l}$ with the addition of one user of class j. Here, it is noteworthy that $\pi(N_l^i)$ and $B_{(j,\,i,\,l)}$ are dependent on the admission region $S_{sub,l}$ and the traffic loads $\rho_{(j,\,i,\,l)}$.

In order for the calls that are not link blocked in each sector to get the services, there should be sufficient CEs in the BS to support them. If there are insufficient CEs in the BS, the calls will be hard blocked. Because all CEs available in the BS are pooled and assigned to any call regardless of sectors, α_j is introduced to consider the traffic load of the jth service group that is offered to the BS from each sector and further defined as $\alpha_j = \sum_{i=1}^{3} \sum_{l=1}^{P} \rho_{(j,i,l)} \cdot \left(1 - b_{(link,j,l,i)}\right)$ for $j = v, d$. For the purpose of evaluating the hard blocking probability in the BS, let $\mathbf{N}_b = (n_v, n_d)$ be the state of the BS given by the number of calls of each service group in the BS. Then, the state probability of $\mathbf{N}_b$ in the BS is given by

$$\pi(\mathbf{N}_b) = \frac{1}{G_b(R_b)} \frac{\alpha_v^{n_v}}{n_v!} \frac{\alpha_d^{n_d}}{n_d!} \tag{12.4}$$

where $G_b(R_b)$ is a normalizing constant for the state probability of the BS that has to be calculated in order to get $\pi(\mathbf{N}_b)$, which is accumulated to 1 and is given as

$$G_b(R_b) = \sum_{\mathbf{N}_b \in S_b(R_b)} \prod_{j=v}^{d} \frac{\alpha_j^{n_j}}{n_j!} \tag{12.5}$$

$$S_b(R_b) = \{\mathbf{N}_b \mid \mathbf{N}_b \mathbf{A}^T \le R_b\} \tag{12.6}$$

where the jth element of $\mathbf{A}$ corresponds to the required amount of CEs to support a user in the j service group, which depends on the modem structure in the BS, and here is set to 1 for all elements of $\mathbf{A}$. $S_b(R_b)$ is a set of admissible states in the BS, $R_b = N$, and N is the total number of CEs available in the BS.

Similarly to the link blocking case, when there are N CEs in the BS, the closed-form equation for the hard blocking probability of the jth service group in the BS is given as following:

$$b_{(hard,j)} = 1 - \frac{G_b\left(R_b - \mathbf{A}e_j\right)}{G_b(R_b)} \tag{12.7}$$

where $G_b(R_b)$ is the normalizing constant calculated on the whole $S_b(R_b)$, while $G_b(R_b - \mathbf{A}e_j)$ is the constant calculated on the $S_b(R_b - \mathbf{A}e_j)$.

Consequently, the problem to evaluate the soft blocking and hard blocking probabilities for users of the jth service group in the lth subsystem at the ith sector is to solve (12.3) and (12.7), which are mutually linked through coupling parameters $\bar{\rho}_{(j,i)}$ and α_j. For the calculation of these blocking probabilities, in this chapter, we adopt an iteration method. We let $b_{(link,j,l,i)}(m)$ and $b_{(hard,\,j)}(m)$ represent the value of $b_{(link,j,l,i)}$ and $b_{(hard,j)}$ at the mth iteration, respectively, and let $b_{(link,j,l,i)}(0)$ and $b_{(hard,\,j)}(0)$ be the initial value for the recursion. At the mth iteration, $b_{(link,j,l,i)}(m)$ is computed using (12.3) with $\bar{\rho}_{(j,l,i)} = \rho_{(j,l,i)} \cdot \left(1 - b_{(hard,j)}(m-1)\right)$, where we intuitively let $\bar{\rho}_{(j,l,i)}$ be $\rho_{(j,l,i)} \cdot \left(1 - b_{(hard,j)}(m-1)\right)$ to consider the effect of the limited number of CEs in the BS on the link blocking probability in the ith sector through the feedback quantity of $b_{(hard,\,j)}(m\text{–}1)$. At the mth iteration, $b_{(hard,\,j)}(m)$ is also computed using (12.7) with the following expression for α_j.

$$\alpha_j = \sum_{i=1}^{3} \sum_{l=1}^{P} \rho_{(j,i,l)} \cdot \left(1 - b_{(link,j,l,i)}(m-1)\right) \tag{12.8}$$

where α_j is intuitively selected to consider the effect of the user limit and the traffic load of each sector on hard blocking. Thus, the iteration procedure takes the following steps:

1. Define $m = 0$, and set $b_{(hard,\,j)}(0) = 0$.
2. Calculate $b_{(link,j,l,i)}(m)$ with $\bar{\rho}_{(j,l,i)} = \rho_{(j,l,i)} \cdot \left(1 - b_{(hard,j)}(m)\right)$ for all i and j.
3. Calculate $b_{(hard,\,j)}(m+1)$ with $\alpha_j = \sum_{i=1}^{3} \rho_{(j,l,i)} \cdot \left(1 - b_{(link,j,l,i)}(m)\right)$.
4. If $\left(\left|b_{(hard,j)}(m+1) - b_{(hard,j)}(m)\right| / b_{(hard,j)}(m+1)\right) < \tau$ (tolerance parameter), then stop the recursion. Otherwise, set $m = m + 1$ and go back to step 2.

From our numerical experiences, it is observed that this recursion always converges within a few iterations (generally less than five). Also, it is noteworthy that even though we select the coupling parameters $\bar{\rho}_{(j,l,i)}$ and α_j somewhat intuitively, the other forms of the coupling parameters may be adopted for a better calculation of link blocking and hard blocking probabilities.

Finally, the call blocking probability of the jth service group in the lth subsystem at the ith sector, $P_{(blocking,\,j,\,l)}$, is given as follows for the convergence values.

$$P_{(blocking,j,l)} = 1 - \left(1 - b_{(link,j,l,i)}\right) \cdot \left(1 - b_{(hard,j)}\right) \tag{12.9}$$

For multiaccess systems supporting multiclass services, Erlang capacity corresponding to the voice-only system needs to be modified in a vector format to consider the performances of voice and data services simultaneously. In this chapter, Erlang capacity is defined as a set of the average offered traffic load of each service group that can be supported while QoS and GoS requirements are satisfied simultaneously. Then, Erlang capacity of the subsystem l per sector, $C_{Erlang,l}$, can be calculated as follows:

$$C_{Erlang,l} \equiv \left\{\left(\hat{\rho}_{(v,l)}, \hat{\rho}_{(d,l)}\right)\right\} = \left\{\left(\rho_{(v,l)}, \rho_{(d,l)}\right) \mid P_{(blocking,v,l)} \leq P_{(B,v)_{req}}, P_{(blocking,d,l)} \leq P_{(B,d)_{req}}\right\} \quad (12.10)$$

where $P_{(B,v)_{req}}$ and $P_{(B,d)_{req}}$ are the required call blocking probabilities of voice and data service groups, respectively, and they can be considered GoS requirements.

Finally, the combined Erlang capacity of the multiaccess system under separate operation, C_{Erlang}, is the sum of those of the subsystems such that

$$C_{Erlang} = \left\{(\rho_v, \rho_d) \mid (\rho_v, \rho_d) \equiv \sum_{l=1}^{P}\left(\rho_{(v,l)}, \rho_{(d,l)}\right),\ \left(\rho_{(v,l)}, \rho_{(d,l)}\right) \in C_{(Erlang,l)} \text{ for } l = 1, \ldots P\right\} \quad (12.11)$$

12.4.2 Erlang Capacity Analysis for Common Operation Method

In the common operation of the multiaccess systems, the admissible region of the multiaccess systems at each sector depends on how users of different types of services are assigned onto the subsystems. That is, according to the employed user assignment scheme in the common operation, the admissible region of multiaccess systems can be one subset of the following set:

$$S_{system} = \left\{\begin{array}{l}\left(n_{(v,i)}, n_{(d,i)}\right) \mid \left(n_{(v,i)}, n_{(d,i)}\right) \equiv \sum_{l-1}^{P}\left(n_{(v,l)}, n_{(d,l)}\right) \\ \text{and } \left(n_{(v,l)}, n_{(d,l)}\right) \in S_{sub,l} \text{ for } l = 1, \ldots, P\end{array}\right\} \quad (12.12)$$

where $n_{(v,i)}$ and $n_{(d,i)}$ are the admissible number of users of voice and data in the multiaccess system at the ith sector.

For the common operation of multiaccess systems, here we consider only two user assignment schemes: a service-based assignment algorithm, which was proposed in [3] as a near-optimum user assignment method, and a rule opposite to the service-based assignment algorithm as the worst-case assignment method. These two cases have a practical meaning because they will provide the upper and lower bound of Erlang capacity of multiaccess system under common operation, respectively.

In the service-based assignment algorithm, we assign users into the subsystem where their expected relative resource cost for the bearer service type in question is the smallest. That is, when a user with service type j is coming in the multiaccess system ($j = v$ or d), then we assign the user to the subsystem $\hat{l}$ that meets the following [3]:

$$\hat{l} = \arg\left\{\min_{l}\left(\frac{\partial f_l\left(n_{(v,l)}, n_{(d,l)}\right)}{\partial n_{(j,l)}} \middle/ \frac{\partial f_l\left(n_{(v,l)}, n_{(d,l)}\right)}{\partial n_{(\sim j,l)}}\right)\right\} \tag{12.13}$$

where $\sim j$ is the "other service" (i.e., if $j = v$, then $\sim j$ is d). For the case that each subsystem has a linear capacity region, then the assignment rule can be simply expressed as $\hat{l} = \arg\left\{\min_{l}\left(\frac{\alpha_{lj}}{\alpha_{l\sim j}}\right)\right\}$.

On the other hand, in the rule opposite to the service-based assignment algorithm, we assign the user having service type j to the subsystem $\hat{l}$ that meets the following:

$$\hat{l} = \arg\left\{\max_{l}\left(\frac{\partial f_l\left(n_{(v,l)}, n_{(d,l)}\right)}{\partial n_{(j,l)}} \middle/ \frac{\partial f_l\left(n_{(v,l)}, n_{(d,l)}\right)}{\partial n_{(\sim j,l)}}\right)\right\} \tag{12.14}$$

According to the employed user assignment scheme, we can obtain the corresponding admissible region of the multiaccess systems under the common operation.

If we denote $S_{s\text{-}based}$ as the admissible region of the multiaccess systems with the service-based assignment scheme, and $S_{opp\text{-}s\text{-}based}$ as one with the rule opposite to the service-based assignment scheme, respectively, then we can calculate corresponding link blocking probability of multiaccess system under the common operation method for these two assignment schemes using the similar method presented in previous section [i.e., by using (12.2) and (12.3) after replacing $S_{sub,l}$ with $S_{s\text{-}based}$ and $S_{opp\text{-}s\text{-}based}$, respectively].

For the hard blocking probability in the common operation method, we can also calculate it using (12.7) after setting coupling parameter α_j as $\sum_{i=1}^{3}\left(1 - b_{soft,j,i}\right)$, where j and i are index for service group and sector, respectively. Here, note that there is no index for subsystems because in the case of the common operation, link capacities of all subsystems are pooled.

Finally, the Erlang capacity of multiaccess system under the common operation method can be calculated by using the iteration method presented in previous section.

12.5 Numerical Results

In this section, we will investigate the Erlang capacity of multiaccess systems with different bearer capacities and quality requirements of subsystems according to the two operation methods (separate and common operation). First, we consider the case there is no CE limitation in the BS (i.e., there are enough CEs in BS). After that, we consider the case that there exists the CE limit in BS.

As a practical example, let's first consider a case with coexisting GSM/EDGE-like and WCDMA-like subsystems. When a spectrum allocation of 5 MHz is assumed for both systems, admissible capacity regions of both systems supporting mixed voice and data traffic are modeled as a linear region such that $f_l(n_{(v,l)}, n_{(d,l)})$ is given as $a_{lv} \cdot n_{(v,l)} + a_{ld} \cdot n_{(d,l)}$ for $l = 1, 2$ where the GSM/EDGE-like system is denoted as subsystem 1, and the WCDMA-like system is denoted as subsystem 2. Furthermore, (a_{1v}, a_{1d}) and (a_{2v}, a_{2d}) are given as (1/62 1/15) and (1/75 1/40), respectively, for standard WCDMA and EDGE data bearers and a circuit switched equivalent bit rate requirement of 150 Kbps [4]. Figure 12.1 shows the resulting Erlang capacity regions when the required call blocking probability is set to 1%.

Lines (i, ii) in Figure 12.1 show the Erlang capacity of GSM/EDGE and WCDMA, respectively. Then, the Erlang capacity of multiaccess systems under the separate operation can be given as the vector sum of those of subsystems, as in the Figure 12.1. It is noteworthy that the Erlang capacity line, stipulating the Erlang capacity region of multiaccess system, depends on the service mix in the subsystems and lies between the minimum bound line—see (iii) in Figure 12.1—and the maximum bound line—see (iv) in Figure 12.1. This means that the shadowed traffic area, delimited by (iii, iv) in Figure 12.1, is not always supported by the multiaccess system under the separate operation. For example, the traffic load of (46, 29) can be supported only when GSM/EDGE supports the voice traffic of 46 and the WCDMA supports the data traffic of 29, but this occasion is very rare. Subsequently, we should operate the system with the Erlang capacity region stipulated by (iii) in Figure 12.1 for the sake of stable system operation.

On the other hand, (v) in Figure 12.1 shows the Erlang capacity region of the multiaccess system under the service-based assignment algorithm. In this case, with the service-based assignment scheme, we assign voice users to GSM/EDGE as far as possible and data users to WCDMA because GSM/EDGE is relatively better at handling voice users than WCDMA, and vice versa for data users. As a result, it is observed that we can get about 60% capacity improvement through the service-based assignment algorithm over the separate operation where we utilize total supportable traffic load of the system for the performance comparison (i.e., the sum of the maximum supportable voice and data traffic load). Line (vi) in Figure 12.1 also

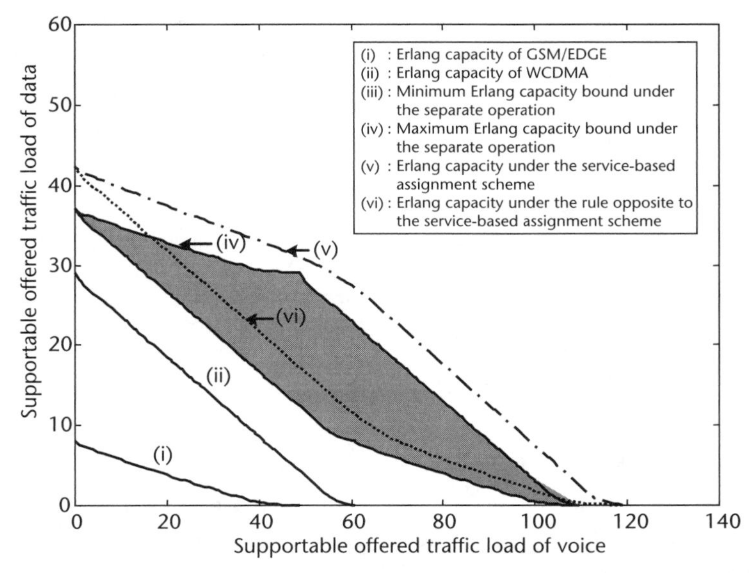

Figure 12.1 Erlang capacity of a GSM/EDGE-like and WCDMA-like multiaccess system.

shows the Erlang capacity region of the multiaccess system when assigning users according to the rule opposite to the service-based assignment algorithm. In this case, the voice users are as far as possible assigned to WCDMA and as many data users as possible are assigned to GSM/EDGE, which corresponds to the worst-case scenario in the common operation. The resulting Erlang capacity is dramatically lower than that of the service-based assignment algorithm. Even in the worst case, however, we know that the common operation still can provide about 15% capacity improvement over the separate operation, in aspect of Erlang capacity.

In addition, we consider an artificial case to consider the effect of air-link capacities of subsystems on the Erlang capacity of multiaccess systems, where the admissible regions of each subsystem are also delimited by the linear bound, and (a_{1v}, a_{1d}) and (a_{2v}, a_{2d}) are given as (1/10 1/10) and (1/20 1/10), respectively.

Figure 12.2 shows the resulting Erlang capacity regions for the two operation methods. With the service-based assignment scheme, in this case we assign voice users to subsystem 2 as far as possible and data users to subsystem 1 because subsystem 2 is relatively better at handling voice users than subsystem 1, and vice versa for data users. As a result, we can achieve a gain of up to 37% over the rule opposite to the service-based assignment through the service-based user assignment, and the gain of up to 88.5% over the separate operation method. When comparing these results with those of the previous example, we also know that the Erlang capacity gains of multiaccess systems, which can be achieved by the operation methods, are very sensitive to subsystem capacities such as the shape and the area of the capacity.

Figure 12.3 shows the Erlang capacity gain of a multiaccess system according to the traffic-mix ratio between voice and data for the previous two numerical examples. Here, we define the traffic-mix ratio as $\rho_v/(\rho_v + \rho_d)$. Noting that the Erlang improvement of common mode operation over the separate operation converges into a trunking gain as the traffic-mix ratio between voice and data goes to 0 or 1, we know that the Erlang improvement of common mode operation is mainly due to the

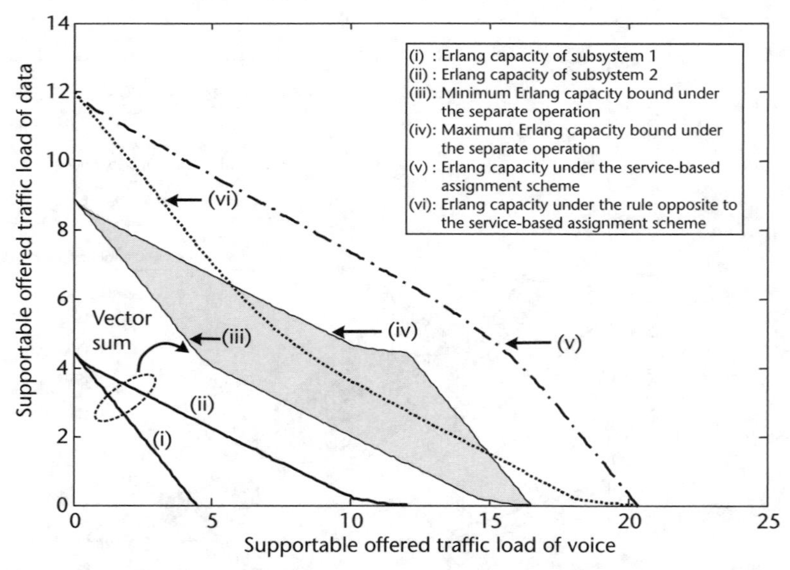

Figure 12.2 Erlang capacity of a multiaccess system for the two operation methods: separate and common operation method.

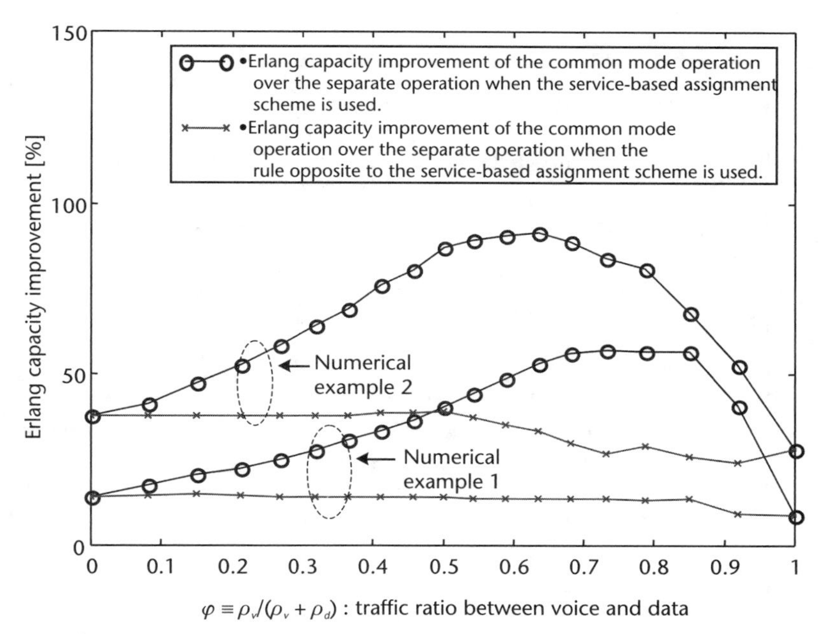

Figure 12.3 Erlang capacity improvement of a multiaccess system according to the traffic-mix ratio between voice and data, φ.

trunking efficiency gain when the rule opposite to the service-based assignment scheme is used. We also know that the gain is less sensitive to the traffic-mix ratio between voice and data, while it is sensitive to the subsystem capacities. On the other hand, Figure 12.3 shows that the Erlang capacity improvement in the case of the service-based assignment scheme varies according to the traffic-mix ratio between voice and data. This means that in this case we can get both a trunking efficiency gain and a service-based assignment gain simultaneously. It is noteworthy that the trunking efficiency gain is rather insensitive to the service mix, whereas the service-based assignment gain depends significantly on the service mix. The service-based assignment scheme is thus more beneficial in mixed-service scenarios.

Until now, we have considered the case that there is no CE limit in BS (i.e., there are enough CEs in the BS). However, multiaccess systems are equipped with a finite number of CEs, afforded in a cost-efficient way because the CEs are a cost part of the system, which inherently affects the Erlang capacity of multiaccess systems.

Figure 12.4 shows the Erlang capacity of the second numerical example case, for different values of CEs. As expected, the Erlang capacities decrease as the number of CEs gets smaller, for both cases of the separate and common operation methods.

However, Erlang capacities under the common operation are more severely affected by the limited number of CEs than those under the separate operation. When the number of CEs is less than 40, the Erlang capacities between the separate and common operations are almost the same, which is mainly because the flexibility of common operation that comes from combining all air-link capacities of subsystems has no influence on the Erlang capacity because call blocking mainly occurs due to insufficient CEs in the BS. However, as the number of CEs available in the BS increases, call blocking gradually occurs due not to insufficient CEs in BS but to air-link capacity limit per sector. Subsequently, the common operation method

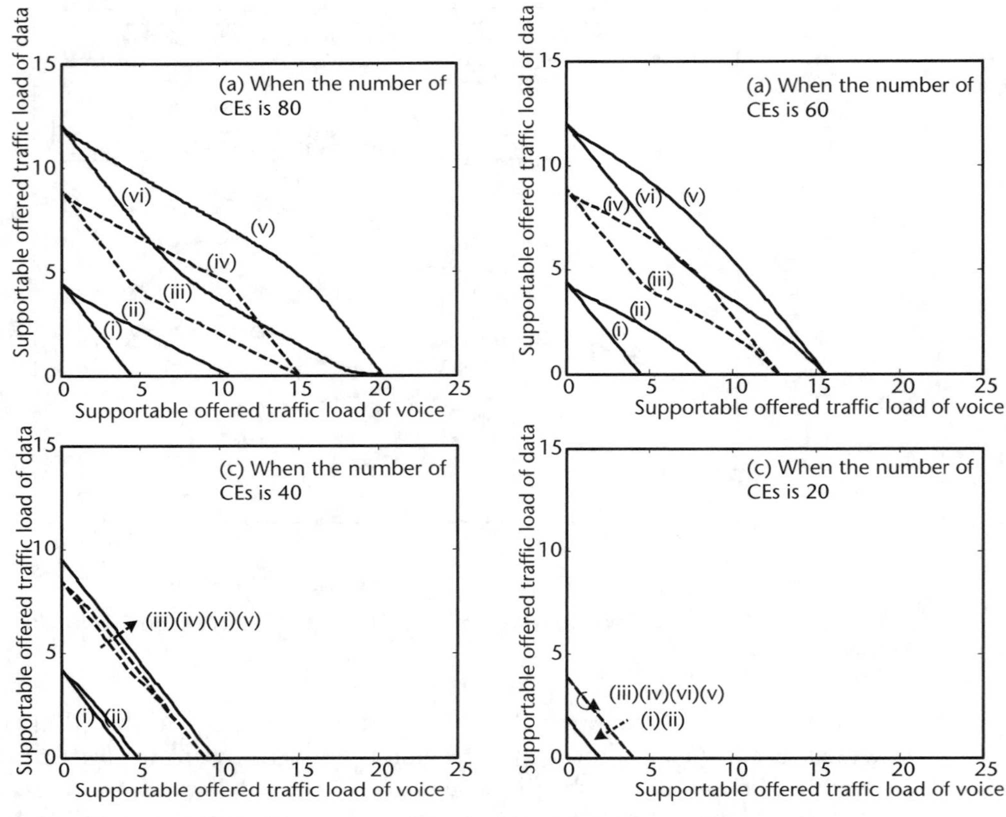

Figure 12.4 Erlang capacity of the multiaccess system for different numbers of CEs.

improves the call blocking probability by pooling the air-link capacities of subsystems and further outperforms the separate operation method for a larger number of CEs.

For a deeper consideration of the effect of CEs on the Erlang capacity of multiaccess systems, let's assume that the offered traffic load of data is proportional to that of voice and further let $\delta(\equiv \rho_d/\rho_v)$ the traffic ratio of data to voice, which allows the observation space of the Erlang capacity to be one dimension. Figure 12.5 shows the Erlang capacities per sector as a function of CEs when δ = 0, 0.3, 0.7, and 1. All solid lines represent Erlang capacities when the service-based assignment algorithm is used as the user assignment scheme under the common operation, while the dotted lines correspond to Erlang capacities under the separate operation, respectively. From Figure 12.5, we observe that the Erlang capacity region can be divided into three regions according to the number of CEs. In the first region, Erlang capacity increases linearly with the increase of CEs. This means that call blocking, in this region, occurs mainly due to the limitation of CEs in the BS.

In the second region, Erlang capacity is determined by the interplay between the limitation of CEs in the BS and the limitation of air-link capacity at each sector. Finally, in the last region, Erlang capacity is saturated where call blocking is mainly due to insufficient air-link capacity per sector, and we cannot get more Erlang capacity by simply equipping more CEs in the BS. Figure 12.5 also shows that Erlang

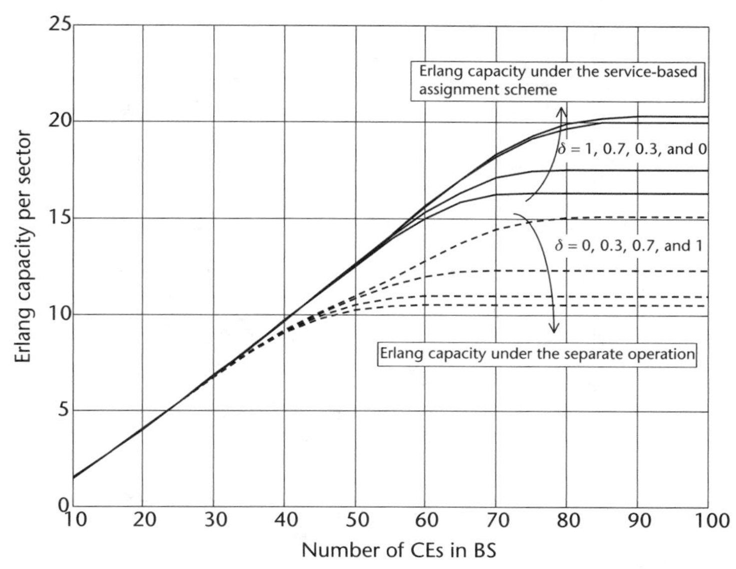

Figure 12.5 Erlang capacity per sector as a function of CEs when $\delta = 0, 0.3, 0.7$, and 1.

capacity of multiaccess systems under the separate operation method is more quickly saturated than that under the common operation method. Practically, it is very important for operators of multiaccess systems to determine or select the proper number of CEs that should be equipped in a BS to fully extract the Erlang capacity of multiaccess systems. With Figure 12.5, we, in this case, can recommend equipping more than 90 CEs in a BS in the case of common operation, and 75 CEs in the case of the separate operation, so as to fully extract corresponding Erlang capacity. In addition, it will result in a waste of hardware resource at the BS to equip more than 90 CEs and 75 CEs in the common and separate operations, respectively. Finally, Figure 12.6 shows corresponding Erlang capacity improvements of the common operation method over the separate operation method when the service-based user assignment is used. As we observed in Figure 12.3, Figure 12.6 also indicates that we can get more gains through the common operation method when the traffic of voice and data calls are properly mixed.

12.6 Conclusion

In this section, we investigate the Erlang capacity of multiaccess systems according to two different operation methods, separate and common operation methods, by simultaneously considering the link capacity limit at each sector as well as the CE limit in the BS. When enough CEs are equipped in the BS, we observe that the Erlang capacity improvement that can be obtained through common operation method is twofold. First, a trunking efficiency gain is achieved due to the combining of resource pools. This gain depends on the subsystem capacities; for small subsystem capacities, the gain is significant. Second, a service-based assignment gain can be achieved by assigning users to the subsystem where their service is most efficiently handled. This gain depends on the shape of the subsystem capacity regions.

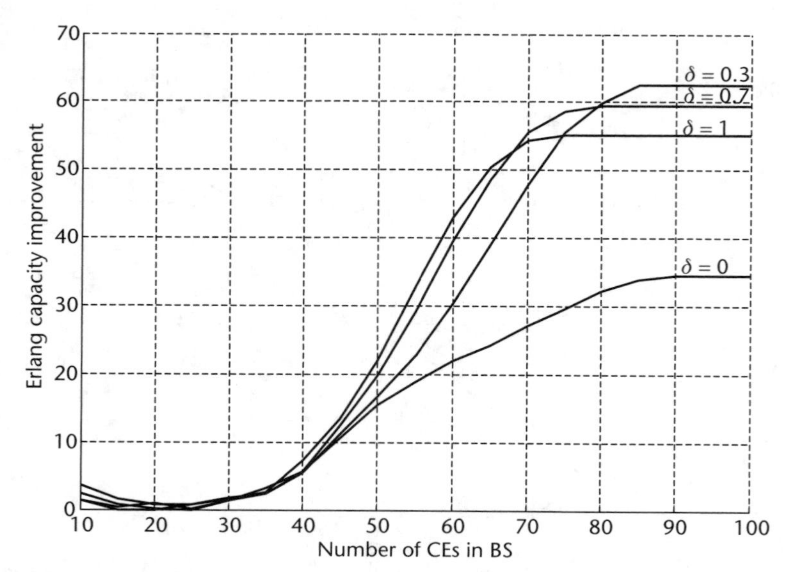

Figure 12.6 Erlang capacity improvements of the common operation method over the separate operation method when the service-based user assignment is used.

Roughly, the more different these are, the larger the gain. It is also observed that the trunking efficiency gain is rather insensitive to the service mix, whereas the service-based assignment gain depends significantly on the service mix. However, the limited number of CEs in the BS reduces the Erlang capacity of multiaccess systems in both cases of common and separate operations. In particular, we know that the Erlang capacities under the common operation are more severely affected by the limited number of CEs than those under the separate operation.

It is subsequently necessary to properly equip CEs in the BS to fully extract the Erlang capacity of multiaccess system while minimizing the equipment cost of the CEs. In the case of the numerical example, we recommend equipping 90 CEs in the BS in the case of the common operation and 75 CEs in the case of the separate operation so as to fully extract corresponding Erlang capacity. Finally, we expect that the results of this chapter would be utilized as a guideline for system operators of multiaccess systems.

References

[1] Ogose, S., "Application of Linkware Radio to the Third Generation Mobile Telecommunications," *IEEE Proc. of VTC*, 1999, pp. 1212–1216.

[2] Tolli, A., P. Hakalin, and H. Holma, "Performance Evaluation of Common Radio Resource Management (CRRM)," *IEEE Proc. of ICC*, 2002, pp. 3429–3433.

[3] Furuskar, A., "Allocation of Multiple Services in Multi-Access Wireless Systems," *IEEE Proc. of MWCN*, 2002, pp. 261–265.

[4] Furuskar, A., "Radio Resource Sharing and Bearer Service Allocation for Multi-Bear Service, Multiaccess Wireless Networks," Ph.D. thesis, 2003, http://www.s3.kth.se/radio/Publication/Pub2003/af_phd_thesis_A.pdf.

[5] Furuskar, A., and J. Zander, "Multi-Service Allocation for Multi-Access Wireless Systems," submitted to *IEEE Transactions on Wireless Communications,* 2002.

[6] Ishikawa, Y., and N. Umeda, "Capacity Design and Performance of Call Admission Control in Cellular CDMA Systems," *IEEE Journal on Selected Areas in Communications*, 1997, pp. 1627–1635.

[7] Yang, J. R., et al., "Capacity Plane of CDMA System for Multimedia Traffic," *IEEE Electronics Letters*, 1997, pp. 1432–1433.

[8] Sampath, A., P. S. Kumar, and J. M. Holtzman, "Power Control and Resource Management for a Multimedia CDMA Wireless System," *IEEE Proc. of International Symposium on Personal, Indoor, and Mobile Radio Communications*, 1995, pp. 21–25.

[9] Koo, I., et al., "A Generalized Capacity Formula for the Multimedia DS-CDMA System," *IEEE Proc. of Asia-Pacific Conference on Communications*, 1997, pp. 46–50.

[10] Kelly, F., "Loss Networks," *The Annals of Applied Probability*, 1991, pp. 319–378.

APPENDIX A

The $M/M/\infty$ Model

Consider a system with Poisson arrivals and exponential service times, and suppose that the number of servers is so large that arriving customers always find a server available. In effect, we have a system with an infinite number of servers. The $M/M/\infty$ system has the transition rate diagram shown in Figure A.1 and further is a birth-death model with

$$\begin{aligned} \lambda_n &= \lambda \quad n = 0, 1, 2, \ldots, \text{and} \\ \mu_n &= n\mu \quad n = 1, 2, \ldots \end{aligned} \tag{A.1}$$

The solution is given by

$$\begin{aligned} P_n &= \prod_{k=0}^{n-1} \frac{\lambda_k}{\mu_{k+1}} = p_0 \prod_{k=0}^{n-1} \frac{\lambda}{(k+1)\mu} \\ &= p_0 \frac{\lambda_n}{\mu(2\mu)\cdots(n\mu)} = p_0 \frac{(\lambda/\mu)^n}{n!}, \quad n = 0, 1, 2, \ldots \end{aligned} \tag{A.2}$$

To find p_0, we use

$$\begin{aligned} 1 &= \sum_{n=0}^{\infty} p_n = \left[\sum_{n=0}^{\infty} \frac{(\lambda/\mu)^n}{n!} \right] p_0 \\ &= e^{\lambda/\mu} p_0 \end{aligned} \tag{A.3}$$

so that

$$p_0 = e^{-\lambda/\mu}$$

and, thus,

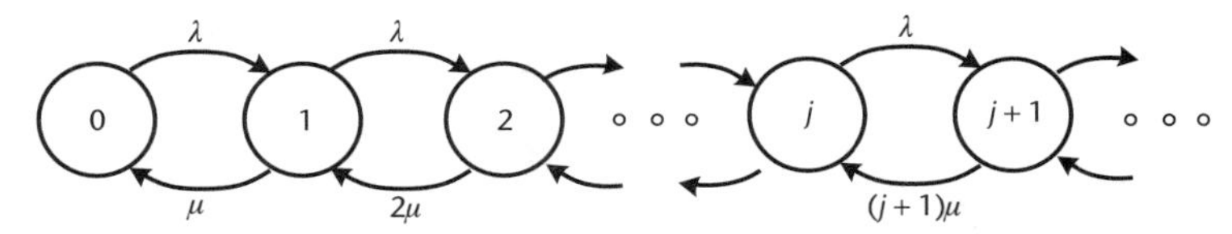

Figure A.1 Transition rate diagram for $M/M/\infty$ model.

$$p_n = \frac{e^{-\lambda/\mu}(\lambda / \mu)^n}{n!}, \quad n = 0, 1, 2, \ldots \tag{A.4}$$

The distribution is Poission with mean λ/μ. The expected number of customers in the system is λ/μ, and the expected response time is $1/\mu = (\lambda/\mu)/\lambda$, the average service time.

APPENDIX B

The *M/M/m* Loss Model

The *M/M/m* loss model has m servers but no waiting room. Calls that arrive when all servers are busy are turned away. This is called a loss system and was first investigated by Erlang. The transition rate diagram for this system is shown in Figure B.1. This is a birth-and-death queuing model with

$$\begin{aligned} \lambda_n &= \lambda, \mu_n = n\mu \quad n = 0, 1, 2, \ldots, m-1 \\ \lambda_n &= 0, \mu_n = m\mu \quad n \geq m \end{aligned} \tag{B.1}$$

The steady state probabilities for this system are given as

$$\begin{aligned} P_n &= \frac{\left(\frac{\lambda}{\mu}\right)^n}{n!} p_0, \quad n = 1, \cdots, m \\ &= 0, \quad n > m \end{aligned} \tag{B.2}$$

and

$$p_0 = \left[\sum_{k=0}^{m} \frac{\left(\frac{\lambda}{\mu}\right)^k}{k!} \right]^{-1} \tag{B.3}$$

Thus,

$$p_n = \frac{\left(\frac{\lambda}{\mu}\right)^n \Big/ n!}{\sum_{k=0}^{m} \left(\frac{\lambda}{\mu}\right)^k \Big/ k!}, n = 0, 1, 2, \ldots, m \tag{B.4}$$

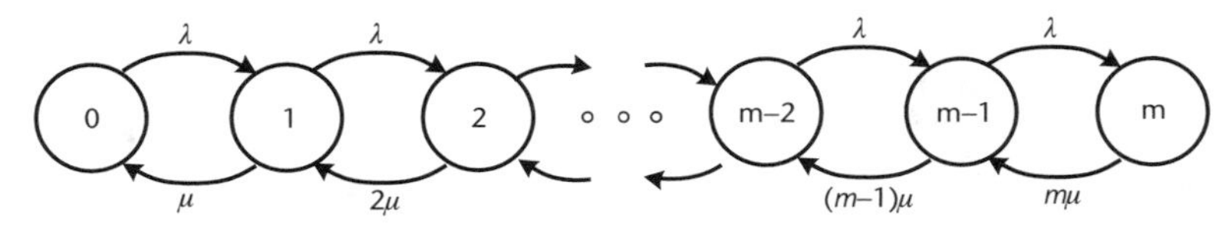

Figure B.1 Transition rate diagram for *M/M/m* loss model.

The distribution of $\{p_n\}$ is truncated Poission. This formula is known as Erlang's first formula. An arriving unit is lost to the system when he finds on arrival that all channels are busy. The probability of this event P_m is

$$P_m = \frac{\left(\frac{\lambda}{\mu}\right)^m \Big/ m!}{\sum_{k=0}^{m} \left(\frac{\lambda}{\mu}\right)^k \Big/ k!} \tag{B.5}$$

Formula (B.5) is known as *Erlang's loss* (or blocking, or overflow) formula, or *Erlang B* formula, and is denoted by $B(m, \lambda/\mu)$. The actual arrival rate into the system is then

$$\lambda_a = \lambda(1 - B(m, \lambda / \mu)) \tag{B.6}$$

The average number in the system is obtained from Little's formula:

$$E[N] = \lambda_a E[\tau] = \frac{\lambda}{\mu}(1 - B(m, \lambda / \mu)) \tag{B.7}$$

Note that the average number in the system is also equal to the carried load. In the case of *M/M/m* loss model, the arrival user will either find an available server or be blocked in the system. If the user finds an available server, then she does not have to wait, and her waiting time in the system equals her service time such that the expected response time is $1/\mu$.

List of Acronyms

1G	First generation
1xEV-DO	High-bit-rate data only
1xEV-DV	High-bit-rate data and voice
2G	Second generation
3G	Third generation
3GPP	Third Generation Partnership Project
AILM	Average interference limited method
AMPS	Advanced mobile phone system
ARQ	Automatic repeat request
ASIC	Application-specific integrated circuit
BER	Bit error rate
BS	Base station
BSC	Base station controller
BTS	Base transceiver subsystem
CAC	Call admission control
CBR	Constant bit rate
CCCA	Combined carrier channel assignment
CDF	Cumulative distribution function
CDMA	Code division multiple access
CE	Channel element
CLSP	Channel load sensing protocol
DS	Direct sequence
DSP	Digital signal processor
DTX	Discontinuous transmission mode
ETC	Equivalent telephone capacity
FA	Frequency allocation
FCFS	First come first served
FDMA	Frequency division multiple access
FFT	Fast Fourier transform
FIFO	First in first out
FPGA	Field-programmable gate array
GoS	Grade of service
GSM	Global system mobile

HSDPA	High-speed downlink packet access
IID	Independent and identically distributed
ICAC	Interference-based CAC
ICCA	Independent carrier channel assignment
IS-95	Interim Standard 95
ISDN	Integrated service digital network
IP	Internet protocol
MAI	Multiaccess interference
MS	Mobile station
MSC	Mobile switching center
MUD	Multiuser detection
NCAC	Number-based CAC
NMT	Nordic mobile telephones
OFDM	Orthogonal frequency division multiplexing
OFDMA	Orthogonal frequency division multiple access
PDN	Public data network
PSTN	Public switched telephone network
QoS	Quality of service
RRM	Radio resource management
SDMA	Space division multiple access
SILM	Statistical interference limited method
SIR	Signal-to-interference ratio
SNR	Signal-to-noise ratio
TACS	Total access communications system
TDMA	Time division multiple access
VBR	Variable bit rate
WCDMA	Wideband CDMA

About the Authors

Kiseon Kim received his B.Eng. and M.Eng. from Seoul National University, both in electronics engineering, in 1978 and 1980, respectively, and his Ph.D. from the University of Southern California, Los Angeles, California, in 1987, in electrical engineering systems.

From 1988 to 1991, he worked for Schlumberger in Texas as a senior development engineer, where he was involved in the development of telemetry systems. From 1991 to 1994, he was a computer communications specialist for Superconducting Super Collider Laboratory in Texas, where built telemetry logging and analysis systems for high-energy physics instrumentations. Since joining Kwang-Ju Institute of Science and Technology (K-JIST), Kwang-Ju, South Korea, in 1994, he has been a professor. His research interests include wideband digital communications system design, analysis, and implementation.

Insoo Koo received a bachelor of engineering in electronic and engineering from Kon-Kuk University, Seoul, South Korea, in 1996 and received his M.E. and Ph.D. from K-JIST in 1998 and 2002, respectively. In 2003, Dr. Koo was a postdoctoral fellow at the Royal Institute of Science and Technology (KTH), Sweden, where he was engaged in the research of packet scheduling algorithms for CDMA-based high-rate packet data systems such as 1xEV-DO, as well as capacity analysis of multiaccess systems. Since 2002, he has worked for the Ultrafast Fiber-Optic Networks (UFON) Research Center, Kwang-Ju Institute of Science and Technology, South Korea, where his research involves the areas of high-speed mobile transmission technologies. At the UFON Research Center, he is currently a research professor. His current research interests include resource management for OFDMA-based high-speed Internet systems.

Index

D

N

O

Recent Titles in the Artech House Mobile Communications Series

John Walker, Series Editor

3G CDMA2000 Wireless System Engineering, Samuel C. Yang

3G Multimedia Network Services, Accounting, and User Profiles, Freddy Ghys, Marcel Mampaey, Michel Smouts, and Arto Vaaraniemi

Advances in 3G Enhanced Technologies for Wireless Communications, Jiangzhou Wang and Tung-Sang Ng, editors

Advances in Mobile Information Systems, John Walker, editor

Advances in Mobile Radio Access Networks, Y. Jay Guo

CDMA for Wireless Personal Communications, Ramjee Prasad

CDMA Mobile Radio Design, John B. Groe and Lawrence E. Larson

CDMA RF System Engineering, Samuel C. Yang

CDMA Systems Capacity Engineering, Kiseon Kim and Insoo Koo

CDMA Systems Engineering Handbook, Jhong S. Lee and Leonard E. Miller

Cell Planning for Wireless Communications, Manuel F. Cátedra and Jesús Pérez-Arriaga

Cellular Communications: Worldwide Market Development, Garry A. Garrard

Cellular Mobile Systems Engineering, Saleh Faruque

The Complete Wireless Communications Professional: A Guide for Engineers and Managers, William Webb

EDGE for Mobile Internet, Emmanuel Seurre, Patrick Savelli, and Pierre-Jean Pietri

Emerging Public Safety Wireless Communication Systems, Robert I. Desourdis, Jr., et al.

The Future of Wireless Communications, William Webb

GPRS for Mobile Internet, Emmanuel Seurre, Patrick Savelli, and Pierre-Jean Pietri

GPRS: Gateway to Third Generation Mobile Networks, Gunnar Heine and Holger Sagkob

GSM and Personal Communications Handbook, Siegmund M. Redl, Matthias K. Weber, and Malcolm W. Oliphant

GSM Networks: Protocols, Terminology, and Implementation, Gunnar Heine

GSM System Engineering, Asha Mehrotra

Handbook of Land-Mobile Radio System Coverage, Garry C. Hess

Handbook of Mobile Radio Networks, Sami Tabbane

High-Speed Wireless ATM and LANs, Benny Bing

Interference Analysis and Reduction for Wireless Systems, Peter Stavroulakis

Introduction to 3G Mobile Communications, Second Edition, Juha Korhonen

Introduction to Digital Professional Mobile Radio, Hans-Peter A. Ketterling

Introduction to GPS: The Global Positioning System, Ahmed El-Rabbany

An Introduction to GSM, Siegmund M. Redl, Matthias K. Weber, and Malcolm W. Oliphant

Introduction to Mobile Communications Engineering, José M. Hernando and F. Pérez-Fontán

Introduction to Radio Propagation for Fixed and Mobile Communications, John Doble

Introduction to Wireless Local Loop, Second Edition: Broadband and Narrowband Systems, William Webb

IS-136 TDMA Technology, Economics, and Services, Lawrence Harte, Adrian Smith, and Charles A. Jacobs

Location Management and Routing in Mobile Wireless Networks, Amitava Mukherjee, Somprakash Bandyopadhyay, and Debashis Saha

Mobile Data Communications Systems, Peter Wong and David Britland

Mobile IP Technology for M-Business, Mark Norris

Mobile Satellite Communications, Shingo Ohmori, Hiromitsu Wakana, and Seiichiro Kawase

Mobile Telecommunications Standards: GSM, UMTS, TETRA, and ERMES, Rudi Bekkers

Mobile Telecommunications: Standards, Regulation, and Applications, Rudi Bekkers and Jan Smits

Multiantenna Digital Radio Transmission, Massimiliano "Max" Martone

Multipath Phenomena in Cellular Networks, Nathan Blaunstein and Jørgen Bach Andersen

Multiuser Detection in CDMA Mobile Terminals, Piero Castoldi

Personal Wireless Communication with DECT and PWT, John Phillips and Gerard Mac Namee

Practical Wireless Data Modem Design, Jonathon Y. C. Cheah

Prime Codes with Applications to CDMA Optical and Wireless Networks, Guu-Chang Yang and Wing C. Kwong

QoS in Integrated 3G Networks, Robert Lloyd-Evans

Radio Engineering for Wireless Communication and Sensor Applications, Antti V. Räisänen and Arto Lehto

Radio Propagation in Cellular Networks, Nathan Blaunstein

Radio Resource Management for Wireless Networks, Jens Zander and Seong-Lyun Kim

RDS: The Radio Data System, Dietmar Kopitz and Bev Marks

Resource Allocation in Hierarchical Cellular Systems, Lauro Ortigoza-Guerrero and A. Hamid Aghvami

RF and Microwave Circuit Design for Wireless Communications, Lawrence E. Larson, editor

Sample Rate Conversion in Software Configurable Radios, Tim Hentschel

Signal Processing Applications in CDMA Communications, Hui Liu

Software Defined Radio for 3G, Paul Burns

Spread Spectrum CDMA Systems for Wireless Communications, Savo G. Glisic and Branka Vucetic

Third Generation Wireless Systems, Volume 1: Post-Shannon Signal Architectures, George M. Calhoun

Traffic Analysis and Design of Wireless IP Networks, Toni Janevski

Transmission Systems Design Handbook for Wireless Networks, Harvey Lehpamer

UMTS and Mobile Computing, Alexander Joseph Huber and Josef Franz Huber

Understanding Cellular Radio, William Webb

Understanding Digital PCS: The TDMA Standard, Cameron Kelly Coursey

Understanding GPS: Principles and Applications, Elliott D. Kaplan, editor

Understanding WAP: Wireless Applications, Devices, and Services, Marcel van der Heijden and Marcus Taylor, editors

Universal Wireless Personal Communications, Ramjee Prasad

WCDMA: Towards IP Mobility and Mobile Internet, Tero Ojanperä and Ramjee Prasad, editors

Wireless Communications in Developing Countries: Cellular and Satellite Systems, Rachael E. Schwartz

Wireless Intelligent Networking, Gerry Christensen, Paul G. Florack, and Robert Duncan

Wireless LAN Standards and Applications, Asunción Santamaría and Francisco J. López-Hernández, editors

Wireless Technician's Handbook, Second Edition, Andrew Miceli

For further information on these and other Artech House titles, including previously considered out-of-print books now available through our In-Print-Forever® (IPF®) program, contact:

Artech House
685 Canton Street
Norwood, MA 02062
Phone: 781-769-9750
Fax: 781-769-6334
e-mail: artech@artechhouse.com

Artech House
46 Gillingham Street
London SW1V 1AH UK
Phone: +44 (0)20 7596-8750
Fax: +44 (0)20 7630-0166
e-mail: artech-uk@artechhouse.com

Find us on the World Wide Web at: www.artechhouse.com

RECEIVED